T0296166

LECTURES

ON

THE LUNAR THEORY.

LECTURES

ON

THE LUNAR THEORY

BY

JOHN COUCH ADAMS, M.A., F.R.S.

LATE LOWNDEAN PROFESSOR OF ASTRONOMY AND GEOMETRY
IN THE UNIVERSITY OF CAMBRIDGE.

EDITED BY

R. A. SAMPSON, M.A.

PROFESSOR OF MATHEMATICS IN THE UNIVERSITY OF DURHAM.

CAMBRIDGE:
AT THE UNIVERSITY PRESS.
1900

CAMBRIDGE
UNIVERSITY PRESS

University Printing House, Cambridge CB2 8BS, United Kingdom

Cambridge University Press is part of the University of Cambridge.

It furthers the University's mission by disseminating knowledge in the pursuit of education, learning and research at the highest international levels of excellence.

www.cambridge.org
Information on this title: www.cambridge.org/9781107559844

First published 1900
First paperback edition 2015

A catalogue record for this publication is available from the British Library

ISBN 978-1-107-55984-4 Paperback

PREFACE.

THE following lectures were collected from manuscripts left by the late Professor J. C. Adams, and are now reprinted without change from his Collected Scientific Papers, Vol. II., pp. 1—84.

It was thought that the wide interest attaching to the lunar problem reached many besides the professed astronomer, and would justify a separate publication of this short work.

It is known that Adams contemplated the publication of some such essay himself, and it must be a matter of regret to all that he never did so. No pains have been spared to present the material properly, but it is unavoidable that it should appear from the hands of an editor in a less perfect form than if the author had issued it himself.

Yet, allowing for this disadvantage, I think those best qualified to judge will consider this work fully worthy of Adams's great name. Of current elementary theories it may be said that they leave off where the difficulties of the subject begin, that is to say, where the various cases of slow convergence have been exposed, but not dealt with. It is perhaps not too much to say that these lectures carry us to the point where such difficulties end, in an adequate evaluation of all the chief constants. They leave the problem effectively solved and not merely stated, and shew the path clear for the formation of a detailed theory, if that is desired.

R. A. SAMPSON.

DURHAM.
8 *October*, 1900.

CONTENTS.

LECTURE		PAGE
I.	HISTORICAL SKETCH	3
II.	ACCELERATIONS OF THE MOON RELATIVE TO THE EARTH .	7
III.	THE SUN'S COORDINATES IN TERMS OF THE TIME . .	13
IV.	THE VARIATION	16
V.	THE VARIATION (*continued*)	20
VI.	THE VARIATION (*continued*)	25
VII.	CORRECTION OF APPROXIMATE SOLUTIONS	30
VIII.	THE PARALLACTIC INEQUALITY	34
IX.	THE PARALLACTIC INEQUALITY (*continued*) . . .	39
X.	THE ANNUAL EQUATION	43
XI.	THE EQUATION OF THE CENTRE AND THE EVECTION .	48
XII.	THE EVECTION AND THE MOTION OF THE APSE . .	54
XIII.	THE MOTION OF THE APSE, AND THE CHANGE OF THE ECCENTRICITY	61
XIV.	THE LATITUDE AND THE MOTION OF THE NODE . .	68
XV.	MOTION IN AN ORBIT OF ANY INCLINATION . . .	72
XVI.	MOTION IN AN ORBIT OF ANY INCLINATION (*continued*) .	76
XVII.	ON HILL'S METHOD OF TREATING THE LUNAR THEORY .	81
XVIII.	ON HILL'S METHOD OF TREATING THE LUNAR THEORY (*cont.*)	85

LECTURES ON THE LUNAR THEORY.

[LECTURES on the Lunar Theory were given by Adams from 1860 with few intermissions until 1889. Originally their aim was to illustrate geometrically the analytical processes and thereby render them more comprehensible, and they included some elegant theorems on the geometry of conics which have since become common property; but every year several lectures were rewritten, and thus the whole fabric gradually changed into the form in which it is here presented,—the form, practically, in which he gave them last.

Perhaps it is superfluous to say that these Lectures stand upon a different footing to treatises that are intended to form the basis of Tables. With such, completeness is the first object and manner of presentation is secondary. Immense as is the labour of forming a treatise of this description, there exist several that leave little to desire in respect to fulness of detail. Indeed it may be suspected that their very perfection in the quality they profess has stifled to some degree the proper development of the subject, because at first sight it suggests that there is little left to do in the Lunar Theory, unless one is prepared to track down the inconsiderable errors that have eluded his Masters. This seems a mistake; the methods most suitable for the whole task adapt themselves comparatively ill to each detail of it, and there seems much that remains to be done in respect to inventing methods suitable for attacking separately, as far as they permit of separate attack, the many difficulties into which the theory divides at the outset, and thence perhaps approximating to a more adequate knowledge than we now possess of the relative motion of Three Bodies. So far, with the notable exception of Dr G. W. Hill and those that have followed him, we have seen comparatively little effort in this direction.

This was the cardinal feature of Adams's plan, and his lectures shew the methods he had gradually elaborated to accomplish it. They separate the inequalities from one another as far as possible, and are content with indicating the manner in which these separate inequalities afterwards combine. To shew that, with so slight an apparatus and within so small a compass, the result is no mere sketch, we need but set side by side the coefficients of longitude found in these Lectures and the corresponding terms in Delaunay's *Théorie*.

		Adams.	Delaunay.
Variation, coeff. of	$\sin 2D$	2106″·4	2106″·25
	$\sin 4D$	8·74	8·75
Parallactic inequality,	$\sin D$	− 124·90*	− 127·62
	$\sin 3D$	0·73	0·84
	$\sin 5D$	0·01	0·01
Annual equation,	$\sin l'$	− 658·9	− 659·23
	$\sin (2D - l')$	152·09	152·11
	$\sin (2D + l')$	− 21·57	− 21·63
Evection,	$\sin (2D - l)$	4596·6	4607·77
	$\sin (2D + l)$	175·1	174·87

Further,

Motion of Apse,	$1 - c$	·008554	·008572
Motion of Node,	$g - 1$	·003997	·003999

For those to whom the difficulties of the Lunar Theory are known, these numbers need no comment.

No Manuscript exists of Lecture I. It is taken substantially from my own notes of 1889.]

* With Delaunay's value of the Sun's Parallax, viz. 8″·75.

LECTURE I.

HISTORICAL SKETCH.

[THE Lunar Theory may be said to have had its commencement with Newton. Many irregularities in the Moon's motion were known before his time, but it was he that first explained the cause of those irregularities and calculated their amounts from theory.

Of the inequalities which are due to the action of the Sun, the first,—which is called the Evection,—was discovered by Ptolemy, who lived at Alexandria in the first half of the second century of our era, under the reigns of Hadrian and Antoninus Pius. At a very early period the relative distance of the Moon at different times could be told from the angle it subtended, and its orbit could thus be mapped out. By such means Ptolemy found that its form was not the same from month to month, and that the longer axis moved continually though not uniformly in one direction. He represented this change by a motion of the centre of the ellipse, as we would put it, in an epicycle round the focus, obtaining thus a variable motion for the longer axis and a variable eccentricity.

The representation of position by means of epicycles is intimately related to the modern method of developing the coordinates in harmonic series; thus if we have

$$x = A_1 \cos(n_1 t + \alpha_1) + A_2 \cos(n_2 t + \alpha_2) + \dots$$
$$y = A_1 \sin(n_1 t + \alpha_1) + A_2 \sin(n_2 t + \alpha_2) + \dots$$

the motion of the point (x, y) is that on a circle of radius A_1 with angular velocity n_1, around a centre which moves on a circle of radius A_2 with angular velocity n_2, and so on; and if, more generally, we have

$$x = A_1 \cos (n_1 t + \alpha_1) + \cdots$$
$$y = B_1 \sin (n_1 t + \alpha_1) + \cdots$$

we may reduce this case to the former by rewriting

$$x = \frac{1}{2}(A_1 + B_1) \cos (n_1 t + \alpha_1) + \frac{1}{2}(A_1 - B_1) \cos (-n_1 t - \alpha_1) + \cdots,$$

$$y = \frac{1}{2}(A_1 + B_1) \sin (n_1 t + \alpha_1) + \frac{1}{2}(A_1 - B_1) \sin (-n_1 t - \alpha_1) + \cdots.$$

Probably we have here the reason why circular motions and epicycles were first employed.

Tycho Brahe (1546—1601) discovered the existence of another inequality in the Moon's Longitude quite different from the Elliptic Inequality and the Evection. He found it bore reference to the position of the Sun with regard to the Moon; so that when the Sun and the Moon were in conjunction or opposition or quadratures the position of the Moon was quite well represented by the existing theory, but from conjunction to the quadrature following, her position was more advanced than the place assigned to it, reaching a maximum of some 35′ about half-way; and in the second quadrant it was just as much behind. This inequality he called the Variation; it was the first that Newton accounted for theoretically, and if we were to suppose the Moon and Sun to move, except for mutual disturbance, in pure circles in the same plane, it is the only one that would present itself.

The next significant step was made by Horrox (1619—1641) who represented the Evection geometrically by motion in a variable ellipse, and gave very approximately the law of variation of the eccentricity and the motion of the apse. He supposed the focus of the orbit to move in an epicycle about its mean place.

Newton's *Principia* did not profess to be and was not intended for a complete exposition of the Lunar Theory. It was fragmentary; its object was to shew that the more prominent irregularities admitted of explanation on his newly discovered theory of universal gravitation. He explained the Variation completely, and traced its effects in Radius Vector as well as in Longitude; and he also saw clearly that the change of eccentricity and motion of the apse that constitute the Evection could be explained on his principles, but he did not give the investigation in the *Principia*, even to the extent to which he had actually carried it. The approximations are more difficult in this case than in that of the Variation, and require to be carried further in order to furnish results of the same accuracy as had already been obtained by Horrox from observation. He was more successful in dealing with the motion of the node and the law of change of inclination. He shewed that when Sun and Node were in conjunction, then for nearly a month the Moon moved in a plane very approximately, and that the inclination of the orbit then reached its maximum, namely, 5° 17′ about; but as the Sun moved away from the Node the latter also began to move, attaining its greatest rate when the separation was a quadrant, and that at this instant the inclination was 5° very nearly. He also assigned the law for intermediate positions. The fact that there was no motion when the Sun was at the Node, that is, in the plane of the Moon's orbit, confirmed his theory that these inequalities were due to the Sun's action.

When we spoke of Newton's results as fragmentary and incomplete, let it not be understood that he gave only very rude approximations to the truth. His results are far more accurate than those arrived at in elementary works of the present day upon the subject.

After Newton, Clairaut (1713—1765) treated the Lunar Theory analytically. He readily found the Variation and many other inequalities, but met with a difficulty in determining the motion of the apse. At first he made its mean motion only about one-half of the observed value, and supposed that this indicated a failure of Newton's law of the inverse square of the distance; but soon he recognized an error, caused by omission of terms which he had imagined would not affect the result. When these were included the calculated amount was nearly doubled.

The first Tables of the Moon which were sufficiently accurate for use in determining longitudes at sea by observation of Lunar Distances were those of Mayer. They obtained a prize offered by our Board of Longitude, and were published in 1770 by Maskelyne, the Astronomer Royal.

The first Theories which gave the Moon's place with an accuracy equal to that of observation were those of Damoiseau and Plana. The former was published in 1827, preceded in 1824 by Tables; the latter was published in 1832.

Hansen's Tables, which are those now used, were constructed from theory and were published in 1857 at the expense of the British Government.]

LECTURE II.

ACCELERATIONS OF THE MOON RELATIVE TO THE EARTH.

WHEN three bodies move under their mutual attraction, their motions are unknown to us except in the cases when they are approximately elliptical; but this restriction includes almost all the most important cases in the Solar System.

If one body of the system is greatly predominant and if the lesser bodies are not close together, the centre of gravity of the greater body may be taken as a common focus around which the others move in approximate ellipses. Or again, if two bodies lie close together, their relative motion may be approximately the same as though they were isolated, although the system contains a third greatly predominant body; for their relative motion is affected by the difference of the attractions of the central body upon them and not by the absolute value of those attractions.

The Sun and Planets are an example of the first kind; the Earth, Moon and Sun of the second. The Earth and Moon describe orbits round the Sun which are approximately ellipses, and the Moon might be regarded as one of the planets; but this point of view would not be a simple one; the disturbing action of the Earth would be too great, though it is never so great as the direct attraction of the Sun, that is to say, never great enough to make the Moon's path convex to the Sun. The more convenient method is to refer the motion of the Moon to the Earth, and counting only the difference of the attractions of the Sun upon the Earth and upon the Moon, to find how this distorts the otherwise elliptical relative orbit. This is the method of the Lunar Theory.

The position of the Sun must be referred to the same origin; but since the Earth describes an ellipse about the Sun which is disturbed by the action of the Moon, if we choose as origin the Earth's centre, we must allow for the disturbance of the Sun's position by the Moon. This correction may be evaded by choosing as origin, not the Earth's centre, but the centre of gravity of the Earth and Moon, with respect to which the Sun describes a curve so closely elliptical that no allowance is required. For, if S, E, M denote respectively the Sun, Earth, and Moon, and G the centre of gravity of E and M, the accelerating forces of S are

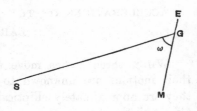

$$\text{on } E \quad S/SE^2 \text{ in } ES,$$
$$\text{on } M \quad S/SM^2 \text{ in } MS;$$

and these imply accelerations of G of amount

$$\frac{E}{E+M}\frac{S}{SE^2} \text{ parallel to } ES,$$

$$\frac{M}{E+M}\frac{S}{SM^2} \text{ parallel to } MS;$$

now the accelerations of S are

$$E/SE^2 \text{ in } SE,$$
$$M/SM^2 \text{ in } SM;$$

hence the acceleration of G relative to S is

$$\frac{S+E+M}{E+M}\frac{E}{SE^2} \text{ parallel to } ES,$$

$$\frac{S+E+M}{E+M}\frac{M}{SM^2} \text{ parallel to } MS;$$

or $\qquad \dfrac{S+E+M}{E+M}\left(E\cdot\dfrac{GE}{SE^3}-M\cdot\dfrac{GM}{SM^3}\right) \text{ in } GM,$

$$\frac{S+E+M}{E+M}\left(E\cdot\frac{SG}{SE^3}+M\cdot\frac{SG}{SM^3}\right) \text{ in } GS.$$

Let $\qquad EM = r,\ SG = r',\ SGM = \omega$; then

$$GM = \frac{E}{E+M}\,r, \qquad GE = \frac{M}{E+M}\,r.$$

Hence

$$\frac{1}{SM^3} = \frac{1}{r'^3}\Bigg[1 + \frac{E}{E+M}\,\frac{r}{r'}\,3\cos\omega$$
$$+ \Big(\frac{E}{E+M}\,\frac{r}{r'}\Big)^2\Big(-\frac{3}{2} + \frac{15}{2}\cos^2\omega\Big) + \ldots\ldots\Bigg],$$

$$\frac{1}{SE^3} = \frac{1}{r'^3}\Bigg[1 - \frac{M}{E+M}\,\frac{r}{r'}\,3\cos\omega$$
$$+ \Big(\frac{M}{E+M}\,\frac{r}{r'}\Big)^2\Big(-\frac{3}{2} + \frac{15}{2}\cos^2\omega\Big) + \ldots\ldots\Bigg];$$

and the accelerations of G are

$$\frac{S+E+M}{r'^2}\Bigg[\quad -\frac{EM}{(E+M)^2}\,\frac{r^2}{r'^2}\,3\cos\omega \qquad\quad + \ldots\ldots\Bigg]\ \text{in } GM,$$

$$\frac{S+E+M}{r'^2}\Bigg[1 + \frac{EM}{(E+M)^2}\,\frac{r^2}{r'^2}\Big(-\frac{3}{2} + \frac{15}{2}\cos^2\omega\Big) + \ldots\ldots\Bigg]\ \text{in } GS.$$

Now r/r' is approximately $\dfrac{1}{400}$; neglecting the square of this quantity, we see that S moves about G in a pure ellipse.

Consider now the accelerations of the Moon relative to the Earth ; subtracting the accelerations of the Earth from those of the Moon, we find

$$\frac{E+M}{ME^2} + S\Big(\frac{MG}{SM^3} + \frac{EG}{SE^3}\Big)\ \text{in } MG,$$

$$S\Big(\frac{SG}{SM^3} - \frac{SG}{SE^3}\Big)\ \text{parallel to } GS;$$

let $E+M = \mu,\ S = m'$; then these become

$$\frac{\mu}{r^2} + \frac{m'r}{r'^3}\Bigg[1 \qquad + \frac{E-M}{E+M}\,\frac{r}{r'}\,3\cos\omega \qquad\quad + \ldots\ldots\Bigg]\ \text{in } ME,$$

$$\frac{m'r}{r'^3}\Bigg[3\cos\omega + \frac{E-M}{E+M}\,\frac{r}{r'}\Big(-\frac{3}{2} + \frac{15}{2}\cos^2\omega\Big) + \ldots\ldots\Bigg]$$

$$\text{parallel to } GS.$$

In the accompanying spherical triangle, let G be the centre of the sphere, SM' the ecliptic, and M' the projection of M.

Let $1/u$ be the projection of ME on the plane of the ecliptic;

θ the longitude of the Moon as seen from the Earth,

θ' the longitude of the Sun as seen from G,

s the tangent of the Moon's latitude MM'.

Then

$$SM = \omega, \qquad\qquad SM' = \theta - \theta',$$
$$r = (1 + s^2)^{\frac{1}{2}} u^{-1}, \qquad \cos \omega = \cos (\theta - \theta')(1 + s^2)^{-\frac{1}{2}},$$

and the accelerations of M relative to E are

$$\frac{\mu u^2}{1 + s^2} + \frac{m'(1 + s^2)^{\frac{1}{2}}}{r'^3 u}\left[1 + \frac{E - M}{E + M}\frac{1}{r'u} 3 \cos (\theta - \theta') + \ldots\ldots \right] \text{ in } ME,$$

$$\frac{m'}{r'^3 u}\left[3 \cos (\theta - \theta') + \frac{E - M}{E + M}\frac{1}{r'u} \right.$$
$$\left. \left(-\frac{3}{2}(1 + s^2) + \frac{15}{2}\cos^2 (\theta - \theta') \right) + \ldots\ldots \right] \text{ parallel to } GS.$$

Call these quantities U and V respectively; then if we resolve parallel to $M'G$, perpendicular to $M'G$ in the plane of the ecliptic, and perpendicular to the plane of the ecliptic, we have the following quantities which we call P, T, S; viz. :—

$$P = \quad U (1 + s^2)^{-\frac{1}{2}} - V \cos (\theta - \theta'),$$
$$T = - V \sin (\theta - \theta'),$$
$$S = \quad Us (1 + s^2)^{-\frac{1}{2}};$$

and also

$$S - Ps = \quad Vs \cos (\theta - \theta').$$

From these we find

$$P = \frac{\mu u^2}{(1 + s^2)^{\frac{3}{2}}} - \frac{m'}{r'^3 u}\left[\frac{1}{2} + \frac{3}{2}\cos 2 (\theta - \theta') + \frac{E - M}{E + M}\frac{1}{r'u} \right.$$
$$\left. \left\{ \left(\frac{9}{8} - \frac{3}{2}s^2 \right) \cos (\theta - \theta') + \frac{15}{8}\cos 3 (\theta - \theta') \right\} + \ldots \right],$$

$$T = -\frac{m'}{r'^3 u}\left[\ \frac{3}{2}\sin 2\,(\theta-\theta') + \frac{E-M}{E+M}\frac{1}{r'u}\right.$$

$$\left.\left\{\left(\frac{3}{8}-\frac{3}{2}s^2\right)\sin(\theta-\theta') + \frac{15}{8}\sin 3\,(\theta-\theta')\right\}+ \ldots\right],$$

$$S - Ps = \frac{m's}{r'^3 u}\left[\frac{3}{2} + \frac{3}{2}\cos 2\,(\theta-\theta') + \frac{E-M}{E+M}\frac{1}{r'u}\right.$$

$$\left.\left\{\left(\frac{33}{8}-\frac{3}{2}s^2\right)\cos(\theta-\theta') + \frac{15}{8}\cos 3\,(\theta-\theta')\right\}+ \ldots\right].$$

Hence with the time as independent variable we have the equations of motion

$$\frac{d^2r}{dt^2} - r\left(\frac{d\theta}{dt}\right)^2 = -P,$$

$$\frac{1}{r}\frac{d}{dt}\left(r^2\frac{d\theta}{dt}\right) = T,$$

$$\frac{d^2}{dt^2}(rs) = -S.$$

Or we may write these with θ as independent variable; let

$$r^2\frac{d\theta}{dt} = H,$$

so that

$$\frac{d\theta}{dt} = Hu^2.$$

Then

$$H\frac{dH}{d\theta} = \frac{T}{u^3},$$

$$\frac{d^2r}{dt^2} = -H^2u^2\frac{d^2u}{d\theta^2} - u^2\frac{du}{d\theta}H\frac{dH}{d\theta},$$

$$r\left(\frac{d\theta}{dt}\right)^2 = H^2u^3,$$

whence

$$H^2u^2\left(\frac{d^2u}{d\theta^2}+u\right) + H\frac{dH}{d\theta}u^2\frac{du}{d\theta} = P;$$

again,

$$\frac{d^2}{dt^2}(rs) = H^2u^2\left(u\frac{d^2s}{d\theta^2} - s\frac{d^2u}{d\theta^2}\right) + H\frac{dH}{d\theta}u^2\left(u\frac{ds}{d\theta} - s\frac{du}{d\theta}\right),$$

whence

$$H^2u^3\left(\frac{d^2s}{d\theta^2}+s\right) + H\frac{dH}{d\theta}u^3\frac{ds}{d\theta} = Ps - S;$$

or the equations of motion may be written

$$H^2 u^2 \left(\frac{d^2 u}{d\theta^2} + u\right) = P - T \frac{du}{u d\theta},$$

$$H \frac{dH}{d\theta} = \frac{T}{u^3},$$

$$H^2 u^3 \left(\frac{d^2 s}{d\theta^2} + s\right) = Ps - S - T \frac{ds}{d\theta}.$$

Our problem is to discuss these equations and to obtain from them expressions for the Moon's position at any time. The integration is best effected by observing what kinds of terms will disappear on substitution in the equations, and then assuming for the desired expressions for the coordinates a series of such terms multiplied by undetermined coefficients. Our procedure will be to discuss one by one the irregularities which can be isolated from one another. This will permit a survey of the entire field without involving needless complexity; but if the Lunar Theory is to be accurate, the combinations of such terms with one another must also be included, and the number of terms employed and the labour of manipulating them becomes very great.

LECTURE III.

To obtain the Moon's coordinates in terms of the time from the equations found in Lecture II., we must substitute in the expressions for the forces the developments of the Sun's co-ordinates which we now proceed to give.

Employing as coordinates r', θ', of the last lecture, we have seen that the Sun's motion may be regarded as purely elliptical, so that

$$\frac{a'}{r'} = \frac{1 + e' \cos(\theta' - \varpi')}{1 - e'^2},$$

$$\theta' - \varpi' = n't - \varpi' + 2e' \sin(n't - \varpi') + \frac{5}{4} e'^2 \sin 2(n't - \varpi') + \ldots$$

in which we have written for convenience $n't$ in place of $n't + \epsilon'$.

The quantities that enter the equations are

$$\left(\frac{a'}{r'}\right)^3,$$

$$\left(\frac{a'}{r'}\right)^3 \frac{\cos}{\sin} 2(\theta - \theta'),$$

$$\left(\frac{a'}{r'}\right)^4 \frac{\cos}{\sin} (\theta - \theta'),$$

$$\left(\frac{a'}{r'}\right)^4 \frac{\cos}{\sin} 3(\theta - \theta').$$

Making the substitutions we find without difficulty

$$\left(\frac{a'}{r'}\right)^3 = 1 + \frac{3}{2} e'^2 + 3e' \cos(n't - \varpi' + \frac{9}{2} e'^2 \cos 2(n't - \varpi') + \ldots.$$

$$\left(\frac{a'}{r'}\right)^3 \frac{\cos}{\sin} 2\,(\theta - \theta') = \left(1 - \frac{5}{2}\,e'^2\right) \frac{\cos}{\sin} 2\,(\theta - n't)$$

$$+ \frac{7}{2}\,e' \frac{\cos}{\sin} \{2\,(\theta - n't) - (n't - \varpi')\}$$

$$- \frac{1}{2}\,e' \frac{\cos}{\sin} \{2\,(\theta - n't) + (n't - \varpi')\}$$

$$+ \frac{17}{2}\,e'^2 \frac{\cos}{\sin} \{2\,(\theta - n't) - 2\,(n't - \varpi')\}$$

$$+ \ldots\ldots$$

$$\left(\frac{a'}{r'}\right)^4 \frac{\cos}{\sin} (\theta - \theta') = (1 + 2e'^2) \frac{\cos}{\sin} (\theta - n't)$$

$$+ 3e' \frac{\cos}{\sin} \{(\theta - n't) - (n't - \varpi')\}$$

$$+ e' \frac{\cos}{\sin} \{(\theta - n't) + (n't - \varpi')\}$$

$$+ \frac{53}{8}\,e'^2 \frac{\cos}{\sin} \{(\theta - n't) - 2\,(n't - \varpi')\}$$

$$+ \frac{11}{8}\,e'^2 \frac{\cos}{\sin} \{(\theta - n't) + 2\,(n't - \varpi')\}$$

$$+ \ldots\ldots$$

$$\left(\frac{a'}{r'}\right)^4 \frac{\cos}{\sin} 3\,(\theta - \theta') = (1 - 6e'^2) \frac{\cos}{\sin} 3\,(\theta - n't)$$

$$+ 5e' \frac{\cos}{\sin} \{3\,(\theta - n't) - (n't - \varpi')\}$$

$$- e' \frac{\cos}{\sin} \{3\,(\theta - n't) + (n't - \varpi')\}$$

$$+ \frac{127}{8}\,e'^2 \frac{\cos}{\sin} \{3\,(\theta - n't) - 2\,(n't - \varpi')\}$$

$$+ \frac{1}{8}\,e'^2 \frac{\cos}{\sin} \{3\,(\theta - n't) + 2\,(n't - \varpi')\}$$

$$+ \ldots\ldots$$

These quantities are to be substituted where they occur in the expressions for the forces found in Lecture II.

Let us now make a few general remarks upon the result of the substitution.

It will be observed that the disturbing forces all involve the coefficient $m'a'^{-3}$. It is very important to notice that the Sun's parallax is not required for the evaluation of this quantity. By Kepler's Third Law it is derivable from observations of the Sun's mean motion alone. Other terms however, namely those with the coefficient m'/a'^4u, involve the Sun's parallax directly; and that constant may be obtained by comparing the observed with the theoretical values of the coefficients of those inequalities, with an accuracy probably greater than that of any other method.

The mean disturbing force is radial, and is equal to

$$-\frac{1}{2}\frac{m'a}{a'^3}\left(1+\frac{3}{2}e'^2\right);$$

or the mean effect of the Sun's disturbance is to diminish the Moon's gravity towards the Earth; and to diminish it more, the greater is the eccentricity of the Sun's orbit. Now e' has been diminishing for ages; hence the Moon's gravity towards the Earth has been increasing, and its average time for accomplishing a revolution about the Earth has been diminishing.

This is one cause of the Secular Acceleration of the Moon's mean motion which Halley derived from the records of ancient eclipses.

It may also be noticed that the coefficient of the chief periodic part of the disturbing force, which involves $1-\frac{5}{2}e'^2$, increases as e' diminishes.

Finally let it be observed that the term with argument

$$2(\theta-n't)+2(n't-\varpi'),$$

which does not involve the Sun's Mean Longitude, is absent from the development of $\left(\dfrac{a'}{r'}\right)^3\dfrac{\cos}{\sin}2(\theta-\theta')$.

LECTURE IV.

THE VARIATION.

THE Variation is the first inequality we shall consider; this is the inequality which is independent of eccentricities and mutual inclination in the orbits of the Sun and Moon.

Let us first take the equations in the first form in which they are given in Lecture II., namely with t as independent variable:

$$\frac{d^2r}{dt^2} - r\left(\frac{d\theta}{dt}\right)^2 = -P,$$

$$\frac{1}{r}\frac{d}{dt}\left(r^2\frac{d\theta}{dt}\right) = T;$$

we omit the equation of motion in latitude, and in the expressions for P, T we suppose $s = 0$; moreover it is possible and convenient to discuss separately the terms that involve the Sun's parallax; let these be omitted and we have

$$\frac{1}{r}\frac{d^2r}{dt^2} - \left(\frac{d\theta}{dt}\right)^2 + \frac{\mu}{r^3} = \frac{1}{2}\frac{m'}{r'^3} + \frac{3}{2}\frac{m'}{r'^3}\cos 2(\theta - \theta'),$$

$$\frac{d^2\theta}{dt^2} + \frac{2}{r}\frac{dr}{dt}\frac{d\theta}{dt} = -\frac{3}{2}\frac{m'}{r'^3}\sin 2(\theta - \theta');$$

and if $e' = 0$, $r' = a'$, $m'/a'^3 = n'^2$, $\theta' = n't + \epsilon'$,

$$\frac{1}{r}\frac{d^2r}{dt^2} - \left(\frac{d\theta}{dt}\right)^2 + \frac{\mu}{r^3} = \frac{1}{2}n'^2 + \frac{3}{2}n'^2\cos 2(\theta - n't - \epsilon'),$$

$$\frac{d^2\theta}{dt^2} + \frac{2}{r}\frac{dr}{dt}\frac{d\theta}{dt} = -\frac{3}{2}n'^2\sin 2(\theta - n't - \epsilon');$$

these are the equations to discuss.

Assume as a first approximation

$$\theta = nt + \epsilon + b_2 \sin \{2 (nt + \epsilon) - 2 (n't + \epsilon')\}$$

$$\equiv nt + \epsilon + b_2 \sin 2\psi, \text{ say};$$

$$\frac{1}{r} = \frac{1}{a} [1 + a_2 \cos 2\psi],$$

and we shall suppose a_2, b_2 so small that in the first instance we may neglect their squares and products.

Substitute in the equations; then

$$4(n-n')^2 a_2 \cos 2\psi - \{n^2 + 4n (n - n') b_2 \cos 2\psi\}$$

$$+ \frac{\mu}{a^3} \{1 + 3a_2 \cos 2\psi\} = \frac{1}{2} n'^2 + \frac{3}{2} n'^2 \cos 2\psi$$

$$- 4 (n - n')^2 b_2 \sin 2\psi + 4 (n - n') na_2 \sin 2\psi = \qquad - \frac{3}{2} n'^2 \sin 2\psi.$$

Hence, equating the coefficients of similar terms, we have

$$\frac{\mu}{a^3} = n^2 + \frac{1}{2} n'^2,$$

which gives the relation between n the Moon's mean motion, and $\frac{1}{a}$, the mean of the reciprocal of the distance; also

$$\left[4 (n - n')^2 + \frac{3\mu}{a^3} \right] a_2 - 4n (n - n') b_2 = \quad \frac{3}{2} n'^2 \quad ......(1),$$

$$- 4 (n - n')^2 b_2 + 4n (n - n') a_2 = -\frac{3}{2} n'^2 \quad ......(2).$$

From (2) $4n (n - n') b_2 - 4n^2 a_2 \qquad = \quad \frac{3}{2} \frac{nn'^2}{n - n'}.$

Add to (1), and substitute for μ/a^3;

$$\left[4 (n - n')^2 - n^2 + \frac{3}{2} n'^2 \right] a_2 = \frac{3}{2} n'^2 \frac{2n - n'}{n - n'},$$

$$a_2 = \frac{3}{2} n'^2 \cdot \frac{2n - n'}{n - n'} \cdot \frac{1}{3n^2 - 8nn' + \frac{11}{2} n'^2},$$

$$b_2 = \frac{n}{n-n'}\, a_2 + \frac{3}{8}\frac{n'^2}{(n-n')^2}$$

$$= \frac{3}{2}\, n'^2 \cdot \frac{n\,(2n-n')}{(n-n')^2} \cdot \frac{1}{3n^2 - 8nn' + \dfrac{11}{2}\,n'^2} + \frac{3}{8}\frac{n'^2}{(n-n')^2}.$$

Calling $\dfrac{n'}{n} = m$, we have

$$a_2 = \frac{3}{2}\, m^2 \cdot \frac{2-m}{1-m} \cdot \frac{1}{3 - 8m + \dfrac{11}{2}\,m^2},$$

$$b_2 = \frac{3}{2}\, m^2 \cdot \frac{2-m}{(1-m)^2} \cdot \frac{1}{3 - 8m + \dfrac{11}{2}\,m^2} + \frac{3}{8}\frac{m^2}{(1-m)^2},$$

or, calling $\dfrac{n'}{n-n'}$, or $\dfrac{m}{1-m} = m_1$, we have

$$a_2 = \frac{3}{2}\, m_1{}^2 \cdot \frac{2 + m_1}{3 - 2m_1 + \dfrac{1}{2}\,m_1{}^2},$$

$$b_2 = \frac{3}{2}\, m_1{}^2 \cdot \frac{(1+m_1)\,(2+m_1)}{3 - 2m_1 + \dfrac{1}{2}\,m_1{}^2} + \frac{3}{8}\,m_1{}^2.$$

These are convenient expressions, and, as it happens, very approximate. If we wish to develope in ascending powers of m or m_1, it appears that the latter development will be the more convergent.

We find by observation $\dfrac{n'}{n} = \cdot 07480$, very nearly.

Hence $a_2 = \cdot 00717,95$,

$b_2 = \cdot 01021,2 = 2106''\cdot 4.$

Hence the ratio of the greatest and least distances will be

$$1 \cdot 00717,95 \; : \; 0 \cdot 99282,05,$$

and the greatest angular deviation from the mean longitude will be

$$35'\, 6''\cdot 4,$$

a very close approximation to the truth.

Also we have found

$$\frac{\mu}{a^3} = n^2 + \frac{1}{2}\,n'^2 = n^2\left(1 + \frac{1}{2}\,m^2\right)$$
$$= n^2 \times 1\cdot00280,$$

which is the relation between the actual mean motion and the actual mean distance (or rather mean reciprocal distance) of the Moon.

Without the Sun's disturbing action, the relation between the mean distance and the mean motion, or rather between the radius of the orbit supposed circular and the uniform rate of angular motion along it would be

$$\frac{\mu}{a^3} = n^2.$$

Hence in the actual orbit, the mean motion for a given mean distance is smaller than it would be without disturbance;

Or, for a given mean motion, the mean distance is smaller than it would be without disturbance.

In fact, the relation between the mean distance and the mean motion is the same as it would be if the sum of the masses of the Earth and Moon were diminished in the ratio of $1\cdot00280$ to 1.

LECTURE V.

THE VARIATION (*continued*).

WE will now proceed to substitute in the differential equations the values of $1/r$ and θ which we have obtained, retaining terms of the order of the squares and products of a_2, b_2 and m^2 or m_1^2.

The values to be thus substituted are

$$\frac{1}{r} = \frac{1}{a}(1 + a_2 \cos 2\psi),$$

$$\theta = nt + \epsilon + b_2 \sin 2\psi,$$

where

$$\psi = nt + \epsilon - (n't + \epsilon'),$$

$$a_2 = \frac{3}{2} m_1^2 \frac{2 + m_1}{3 - 2m_1 + \frac{1}{2} m_1^2},$$

$$b_2 = (1 + m_1) a_2 + \frac{3}{8} m_1^2.$$

Hence

$$r = a \left[1 - a_2 \cos 2\psi + \frac{1}{2} a_2^2 (1 + \cos 4\psi) \right],$$

$$\frac{d^2 r}{dt^2} = 4a (n - n')^2 [a_2 \cos 2\psi - 2a_2^2 \cos 4\psi],$$

$$\frac{1}{r} \frac{d^2 r}{dt^2} = 4 (n - n')^2 \left[\frac{1}{2} a_2^2 + a_2 \cos 2\psi - \frac{3}{2} a_2^2 \cos 4\psi \right];$$

again,

$$\frac{d\theta}{dt} = n + 2(n - n') b_2 \cos 2\psi,$$

$$\left(\frac{d\theta}{dt}\right)^2 = n^2 + 4n(n - n') b_2 \cos 2\psi + 2(n - n')^2 b_2^2 [1 + \cos 4\psi],$$

$$\frac{1}{r^3} = \frac{1}{a^3}\left[1 + 3a_2 \cos 2\psi + \frac{3}{2} a_2^2 (1 + \cos 4\psi)\right].$$

Also,

$$\frac{1}{r}\frac{dr}{dt} = 2(n - n')\left[a_2 \sin 2\psi - \frac{1}{2} a_2^2 \sin 4\psi\right],$$

$$\frac{1}{r}\frac{dr}{dt}\frac{d\theta}{dt} = 2(n-n')\left[na_2 \sin 2\psi + \left\{(n - n') a_2 b_2 - \frac{1}{2} na_2^2\right\}\sin 4\psi\right],$$

$$\frac{d^2\theta}{dt^2} = -4(n - n')^2 b_2 \sin 2\psi.$$

And $$\cos 2(\theta - n't - \epsilon') = \cos 2\psi - b_2(1 - \cos 4\psi),$$

$$\sin 2(\theta - n't - \epsilon') = \sin 2\psi + b_2 \sin 4\psi.$$

Substitute these in the differential equations, and we get, on transposing all the terms to the left-hand sides from the first equation

$$4(n - n')^2\left[\frac{1}{2} a_2^2 + a_2 \cos 2\psi - \frac{3}{2} a_2^2 \cos 4\psi\right]$$

$$- [n^2 + 2(n-n')^2 b_2^2 + 4n(n-n') b_2 \cos 2\psi + 2(n-n')^2 b_2^2 \cos 4\psi]$$

$$+ \frac{\mu}{a^3}\left[1 + \frac{3}{2} a_2^2 + 3a_2 \cos 2\psi + \frac{3}{2} a_2^2 \cos 4\psi\right]$$

$$- \frac{1}{2} n'^2 - \frac{3}{2} n'^2 [-b_2 + \cos 2\psi + b_2 \cos 4\psi],$$

and from the second equation

$$-4(n - n')^2 b_2 \sin 2\psi$$

$$+ 4(n - n')\left[na_2 \sin 2\psi + \left\{(n - n') a_2 b_2 - \frac{1}{2} na_2^2\right\}\sin 4\psi\right]$$

$$+ \frac{3}{2} n'^2 [\sin 2\psi + b_2 \sin 4\psi].$$

The coefficient of cos 2ψ in the first of these expressions, and that of sin 2ψ in the second, are respectively

$$4\,(n-n')^2\,a_2 - 4n\,(n-n')\,b_2 + 3\,\frac{\mu}{a^3}\,a_2 - \frac{3}{2}\,n'^2,$$

and $$-4\,(n-n')^2\,b_2 + 4n\,(n-n')\,a_2 \qquad\qquad +\frac{3}{2}\,n'^2,$$

and these are evidently reduced to zero by giving a_2, b_2 the values previously found, if we substitute for μ/a^3 the approximate value $n^2 + \frac{1}{2}\,n'^2$. To find the more correct value of μ/a^3, equate to zero the constant term in the first expression;

$$2(n-n')^2 a_2{}^2 - n^2 - 2(n-n')^2 b_2{}^2 + \frac{\mu}{a^3}\left(1 + \frac{3}{2}\,a_2{}^2\right) - \frac{1}{2}\,n'^2 + \frac{3}{2}\,n'^2 b_2 = 0,$$

that is

$$\frac{\mu}{a^3}\left(1 + \frac{3}{2}\,a_2{}^2\right) = n^2 + \frac{1}{2}\,n'^2 - 2\,(n-n')^2\,a_2{}^2 + 2\,(n-n')^2\,b_2{}^2 - \frac{3}{2}\,n'^2 b_2$$

$$= n^2 + \frac{1}{2}\,n'^2 + 2\,(n-n')^2\left[(2m_1 + m_1{}^2)\,a_2{}^2 - \frac{9}{64}\,m_1{}^4\right].$$

Hence we see that μ/a^3 differs from $n^2 + \frac{1}{2}\,n'^2$ only in terms of the fourth order, if we consider m_1 a quantity of the first order and consequently a_2, b_2 quantities of the second order. Hence also by taking

$$\frac{\mu}{a^3} = n^2 + \frac{1}{2}\,n'^2,$$

in the multiplier of a_2, when we equate to zero the coefficient of cos 2ψ, we only neglected a quantity of the sixth order in m_1, and the error in the resulting values of a_2, b_2 is of that order.

We see that the substitution just made in our equations leaves outstanding terms of the fourth order in cos 4ψ and sin 4ψ. In order to get rid of these we must add terms of this form to the assumed values of $1/r$ and θ, respectively. Suppose that

$$\frac{1}{r} = \frac{1}{a}[1 + a_2 \cos 2\psi + a_4 \cos 4\psi],$$

$$\theta = nt + \epsilon + b_2 \sin 2\psi + b_4 \sin 4\psi,$$

where, as we shall find, a_4 and b_4 are small quantities of the fourth order.

It may be readily seen that the additional terms introduced are the following :—

$$\text{in} \quad \frac{1}{r}\frac{d^2r}{dt^2} \qquad 16\,(n-n')^2\,a_4\cos 4\psi,$$

$$-\left(\frac{d\theta}{dt}\right)^2 \qquad -8n\,(n-n')\,b_4\cos 4\psi,$$

$$\frac{\mu}{r^3} \qquad \frac{\mu}{a^3}\,3a_4\cos 4\psi,$$

$$\frac{d^2\theta}{dt^2} \qquad -16\,(n-n')^2\,b_4\sin 4\psi,$$

$$\frac{2}{r}\frac{dr}{dt}\frac{d\theta}{dt} \qquad 8n\,(n-n')\,a_4\sin 4\psi,$$

and also that we may neglect the terms added in the expressions for

$$\frac{3}{2}\,n'^2\cos 2\,(\theta-n't-\epsilon'), \qquad \frac{3}{2}\,n'^2\sin 2\,(\theta-n't-\epsilon').$$

If we write the terms thus produced along with the several terms left outstanding in our equations and then equate the whole to zero, we have

$$16\,(n-n')^2\,a_4-8n\,(n-n')\,b_4+\frac{\mu}{a^3}3a_4-6\,(n-n')^2\,a_2{}^2-2\,(n-n')^2\,b_2{}^2$$

$$+\frac{\mu}{a^3}\frac{3}{2}\,a_2{}^2-\frac{3}{2}\,n'^2b_2=0,$$

$$-16\,(n-n')^2\,b_4+8n\,(n-n')\,a_4+4\,(n-n')^2\,a_2b_2-2n\,(n-n')\,a_2{}^2$$

$$+\frac{3}{2}\,n'^2b_2=0,$$

from which we must determine a_4 and b_4.

$$\text{Put} \qquad \frac{\mu}{a^3}=n^2+\frac{1}{2}\,n'^2,$$

$$b_2=(1+m_1)\,a^2+\frac{3}{8}\,m_1{}^2,$$

and divide both equations by $(n - n')^2$; we get

$$\left[16 + 3\left(1 + 2m_1 + \frac{3}{2}m_1^2 \right) \right] a_4 - 8\left(1 + m_1 \right) b_4 - \left[6 + 2\left(1 + m_1 \right)^2 \right.$$

$$\left. - \frac{3}{2}\left(1 + 2m_1 + \frac{3}{2}m_1^2 \right) \right] a_2^2 - 3\left(1 + m_1 \right)m_1^2 a_2 - \frac{27}{32}m_1^4 = 0,$$

$$8\left(1 + m_1 \right) a_4 - 16 b_4 + \left[4\left(1 + m_1 \right) - 2\left(1 + m_1 \right) \right] a_2^2$$

$$+ 3m_1^2\left(1 + \frac{1}{2}m_1 \right)a_2 + \frac{9}{16}m_1^4 = 0.$$

Simplify and multiply the last equation by $\frac{1}{2}(1 + m_1)$,

$$\left(19 + 6m_1 + \frac{9}{2}m_1^2 \right) a_4 - 8\left(1 + m_1 \right) b_4 - \left(\frac{13}{2} + m_1 - \frac{1}{4}m_1^2 \right) a_2^2$$

$$- 3\left(1 + m_1 \right)m_1^2 a_2 - \frac{27}{32}m_1^4 = 0,$$

$$\left(4 + 8m_1 + 4m_1^2 \right) a_4 - 8\left(1 + m_1 \right) b_4 + \left(1 + 2m_1 + m_1^2 \right) a_2^2$$

$$+ \frac{3}{2}\left(1 + m_1 \right)\left(1 + \frac{1}{2}m_1 \right)m_1^2 a_2 + \frac{9}{32}\left(1 + m_1 \right)m_1^4 = 0.$$

Subtract the latter from the former and b_4 will be eliminated; we get

$$\left(15 - 2m_1 + \frac{1}{2}m_1^2 \right)a_4 - \left(\frac{15}{2} + 3m_1 + \frac{3}{4}m_1^2 \right) a_2^2$$

$$- \frac{3}{2}\left(1 + m_1 \right)\left(3 + \frac{1}{2}m_1 \right)m_1^2 a_2 - \frac{9}{32}\left(4 + m_1 \right)m_1^4 = 0,$$

which gives a_4; and this being known b_4 is found from

$$b_4 = \frac{1}{2}(1 + m_1)a_4 + \frac{1}{8}(1 + m_1)a_2^2 + \frac{3}{16}\left(1 + \frac{1}{2}m_1 \right)m_1^2 a_2 + \frac{9}{256}m_1^4.$$

Taking $m = \cdot 07480$ as in Lecture IV, we find

$$a_4 = \cdot 00004,580,$$

$$b_4 = \cdot 00004,237 = 8''\cdot 740.$$

LECTURE VI.

THE VARIATION (*continued*).

LET us consider the problem of the Variation over again, taking now θ as independent variable.

The equations of motion are given in Lecture II:—

$$\frac{d^2u}{d\theta^2} + u = \frac{P}{H^2u^2} - \frac{T}{H^2u^3}\frac{du}{d\theta},$$

$$H\frac{dH}{d\theta} = \frac{T}{u^3},$$

where

$$\frac{P}{u^2} = \mu - \frac{1}{2}\frac{n'^2}{u^3} - \frac{3}{2}\frac{n'^2}{u^3}\cos 2(\theta - \theta'),$$

$$\frac{T}{u^3} = \qquad\qquad -\frac{3}{2}\frac{n'^2}{u^4}\sin 2(\theta - \theta'),$$

so that the second equation may be written

$$\frac{1}{H^2}\frac{d(H^2)}{d\theta} = -3n'^2\left(\frac{dt}{d\theta}\right)^2\sin 2(\theta - \theta').$$

Our aim is to express t and u in terms of θ and constant quantities. Now since the orbit of the Moon does not differ widely from a circle we may write the difference of $nt + \epsilon$ from θ, and the difference of au from unity as series of small periodic terms depending upon θ. Inspecting the form of the equations, it is evident that these periodic terms are of argument $2(\theta - \theta')$ and its multiples; that is

$$nt + \epsilon = \theta + \text{periodic terms of argument } 2(\theta - \theta'), \text{ \&c.};$$

but

$$n't + \epsilon' = \theta';$$

therefore

$$\theta - \theta' = (1 - m)\,\theta - \beta$$
$$+ \text{periodic terms of argument } 2\,(1 - m)\,\theta - 2\beta, \text{ &c.},$$

where we have written

$$\beta = \epsilon' - m\epsilon\,;$$

this constant β is associated with $(1 - m)\,\theta$ wherever the latter occurs; for brevity in writing, we shall omit it.

We may then assume as a first approximation

$$au = 1 + a_2 \cos(2 - 2m)\,\theta,$$
$$nt + \epsilon = \theta + b_2 \sin(2 - 2m)\,\theta\,;$$

whence

$$2\,(\theta - \theta') = (2 - 2m)\,\theta - 2mb_2 \sin(2 - 2m)\,\theta,$$
$$\cos 2\,(\theta - \theta') = mb_2 + \cos(2 - 2m)\,\theta - mb_2 \cos(4 - 4m)\,\theta,$$
$$\sin 2\,(\theta - \theta') = \qquad \sin(2 - 2m)\,\theta - mb_2 \sin(4 - 4m)\,\theta,$$
$$n\frac{dt}{d\theta} = 1 + (2 - 2m)\,b_2 \cos(2 - 2m)\,\theta.$$

Substitute in the right-hand member of the second equation :—

$$\frac{1}{H^2}\frac{d(H^2)}{d\theta} = -3m^2\,[\sin(2 - 2m)\,\theta + (2 - 3m)\,b_2 \sin(4 - 4m)\,\theta].$$

Therefore

$$\log_e\left(\frac{H^2}{h^2}\right) = \frac{3}{2}\frac{m^2}{1 - m}\cos(2 - 2m)\,\theta + \frac{3}{4}\frac{2 - 3m}{1 - m}\,m^2 b_2 \cos(4 - 4m)\,\theta,$$

which we may write

$$\log_e\left(\frac{H^2}{h^2}\right) = 2h_2 \cos(2 - 2m)\,\theta + 2h_4 \cos(4 - 4m)\,\theta,$$

where h is an arbitrary constant of integration, h_2 is a known quantity, and h_4 involves b_2. If we take as a second approximation

$$au = 1 + a_2 \cos(2 - 2m)\,\theta + a_4 \cos(4 - 4m)\,\theta,$$
$$nt + \epsilon = \theta + b_2 \sin(2 - 2m)\,\theta + b_4 \sin(4 - 4m)\,\theta,$$

the above value of $\log_e(H^2/h^2)$ will not require modification and will supply equations of condition for determining the coefficients a_2, b_2, a_4, b_4.

Thus
$$n\frac{dt}{d\theta} = \frac{n}{Hu^2} = \frac{na^2}{h}\frac{h}{H}\frac{1}{(au)^2},$$
so that
$$\log_e\left(n\frac{dt}{d\theta}\right) = \log_e\left(\frac{na^2}{h}\right) - \frac{1}{2}\log_e\left(\frac{H^2}{h^2}\right) - 2\log_e au;$$
but
$$\log_e\left(n\frac{dt}{d\theta}\right) = -(1-m)^2 b_2^2 + (2-2m)\,b_2\cos(2-2m)\,\theta$$
$$+ [(4-4m)\,b_4 - (1-m)^2 b_2^2]\cos(4-4m)\,\theta,$$
$$\log_e au = -\frac{a_2^2}{4} + a_2\cos(2-2m)\,\theta + \left(a_4 - \frac{a_2^2}{4}\right)\cos(4-4m)\,\theta.$$

Hence we find
$$-(1-m)^2 b_2^2 = \log_e\left(\frac{na^2}{h}\right) + \frac{1}{2}a_2^2,$$
$$(2-2m)\,b_2 = -h_2 - 2a_2,$$
$$(4-4m)\,b_4 - (1-m)^2 b_2^2 = -h_4 - 2a_4 + \frac{1}{2}a_2^2.$$

The remaining equations of condition that we require are obtained from the first equation of motion; this may be written
$$\frac{d^2(au)}{d\theta^2} + au\left[1 + \frac{1}{2}\left(n'\frac{dt}{d\theta}\right)^2 + \frac{3}{2}\left(n'\frac{dt}{d\theta}\right)^2\cos 2\,(\theta-\theta')\right]$$
$$- \frac{d(au)}{d\theta}\frac{3}{2}\left(n'\frac{dt}{d\theta}\right)^2\sin 2\,(\theta-\theta') = \frac{\mu a}{H^2}.$$

Now
$$au = 1 + a_2\cos(2-2m)\,\theta + a_4\cos(4-4m)\,\theta,$$
whence
$$\frac{d(au)}{d\theta} = -(2-2m)\,a_2\sin(2-2m)\,\theta - (4-4m)\,a_4\sin(4-4m)\,\theta,$$
$$\frac{d^2(au)}{d\theta^2} = -(2-2m)^2 a_2\cos(2-2m)\,\theta - (4-4m)^2 a_4\cos(4-4m)\,\theta,$$
and
$$\left(n'\frac{dt}{d\theta}\right)^2 = m^2[1 + (4-4m)\,b_2\cos(2-2m)\,\theta],$$
$$\left(n'\frac{dt}{d\theta}\right)^2\sin 2(\theta-\theta') = m^2[\sin(2-2m)\,\theta + (2-3m)\,b_2\sin(4-4m)\,\theta],$$

$$\left(n'\frac{dt}{d\theta}\right)^2 \cos 2(\theta - \theta')$$
$$= m^2 \left[(2-m) b_2 + \cos(2-2m)\theta + (2-3m) b_2 \cos(4-4m)\theta\right],$$
$$\frac{1}{H^2} = \frac{1}{h^2} \left[1 + h_2{}^2 - 2h_2 \cos(2-2m)\theta + (h_2{}^2 - 2h_4)\cos(4-4m)\theta\right].$$

Substitute these in the equation above, and equate the coefficients of corresponding terms,

$$1 + \frac{1}{2}m^2 + \frac{3}{2}m^2(2-m) b_2 + \frac{3}{4}m^2 a_2 + \frac{3}{4}m^2(2-2m) a_2 = \frac{\mu a}{h^2}(1 + h_2{}^2)$$

$$-(2-2m)^2 a_2 + \left(1 + \frac{1}{2}m^2\right) a_2 + \frac{3}{2}m^2 + (2-2m) m^2 b_2 = \frac{\mu a}{h^2}(-2h_2)$$

$$-(4-4m)^2 a_4 + \left(1 + \frac{1}{2}m^2\right) a_4 + \frac{3}{2}m^2(2-3m) b_2 + \frac{3}{4}m^2 a_2$$
$$-\frac{3}{4}m^2(2-2m) a_2 = \frac{\mu a}{h^2}(h_2{}^2 - 2h_4).$$

If we neglect at first terms of the fourth order, we find from the first of these equations

$$\frac{\mu a}{h^2} = 1 + \frac{1}{2}m^2.$$

From the earlier set of equations we have

$$(2-2m) b_2 = -h_2 - 2a_2;$$

substitute this in the second equation above. We get

$$\left[-(2-2m)^2 + 1 + \frac{1}{2}m^2 - 2m^2\right] a_2 - m^2 h_2 + \frac{3}{2}m^2 = \left(1 + \frac{1}{2}m^2\right)(-2h_2),$$

or

$$\left(3 - 8m + \frac{11}{2}m^2\right) a_2 = \frac{3}{2}m^2 + 2h_2 = \frac{3}{2}m^2 + \frac{3}{2}\frac{m^2}{1-m} = \frac{3}{2}m^2\frac{2-m}{1-m},$$

so that

$$a_2 = \frac{3}{2}m^2\frac{2-m}{1-m}\frac{1}{3 - 8m + \frac{11}{2}m^2},$$

and

$$b_2 = -\frac{1}{1-m}a_2 - \frac{1}{2-2m}h_2$$
$$= -\frac{1}{1-m}a_2 - \frac{3}{8}\frac{m^2}{(1-m)^2}.$$

These are numerically equal to the quantities denoted by the same symbols in Lecture IV, but b_2 bears the contrary sign. We further find

$$\left(15 - 32m + \frac{31}{2}m^2\right)a_4$$

$$= \frac{3}{4}\frac{m^2}{1-m}\left\{(1 + 3m - 2m^2)a_2 + (8 - 15m + 6m^2)b_2\right\},$$

$$b_4 = -\frac{1}{2-2m}a_4 - \frac{3}{8}a_2 b_2 - \frac{3}{64}\frac{m^2}{(1-m)^2}\left\{a_2 + (6 - 8m)b_2\right\};$$

or reduced to numbers

$$a_4 = -\cdot00002{,}210,$$
$$b_4 = \cdot00005{,}414 = 11''\cdot17.$$

Finally let us exhibit the relation between the constants employed in this investigation and those of Lectures IV, V; to distinguish them, attach accents to the latter, so that

$$\theta = nt + \epsilon + b_2'\sin 2\psi + b_4'\sin 4\psi,$$
$$\frac{a'}{r} = 1 + a_2'\cos 2\psi + a_4'\cos 4\psi,$$

and, omitting the constant β as before,

$$(1-m)\theta = \psi + (1-m)b_2'\sin 2\psi + (1-m)b_4'\sin 4\psi.$$

Then

$$2\psi = (2-2m)\theta - (2-2m)b_2'\sin(2-2m)\theta,$$
$$\sin 2\psi = \sin(2-2m)\theta - (1-m)b_2'\sin(4-4m)\theta,$$
$$\cos 2\psi = (1-m)b_2' + \cos(2-2m)\theta - (1-m)b_2'\cos(4-4m)\theta.$$

Substitute in the equation for θ; we find

$$nt + \epsilon = \theta - b_2'\sin(2-2m)\theta - [b_4' - (1-m)b_2'^2]\sin(4-4m)\theta,$$

and similarly

$$a'u = 1 + (1-m)a_2'b_2' + a_2'\cos(2-2m)\theta$$
$$+ [a_4' - (1-m)a_2'b_2']\cos(4-4m)\theta.$$

We observe that a' differs from a by quantities of the fourth order.

LECTURE VII.

CORRECTION OF APPROXIMATE SOLUTIONS.

We may simplify the equations we have been dealing with, by a proper choice of units. Let the unit of distance be the radius of the circular orbit which the Moon, if undisturbed, would describe about the Earth in its actual periodic time; then

$$\mu = n^2.$$

Also choose the unit of time so that

$$n - n' = 1,$$

so that, if we take as the result of observation of the mean motions of the Sun and Moon,

$$n' : n = \cdot 07480,13,$$

we get

$$n' = \cdot 08084,9 = m_1,$$

where m_1 is the quantity so called in Lecture IV; and

$$\mu = 1 \cdot 16823,4.$$

We shall frequently adopt these simplifications in what follows.

Now let

$$l = \log_e (r/a),$$

so that

$$\frac{1}{r}\frac{dr}{dt} = \frac{dl}{dt}, \quad \frac{1}{r}\frac{d^2 r}{dt^2} = \frac{d^2 l}{dt^2} + \left(\frac{dl}{dt}\right)^2, \quad \frac{\mu}{r^3} = \frac{\mu}{a^3} e^{-3l};$$

and the equations discussed in Lecture IV become

$$\frac{d^2 l}{dt^2} + \left(\frac{dl}{dt}\right)^2 - \left(\frac{d\theta}{dt}\right)^2 + \frac{\mu}{a^3} e^{-3l} - n'^2 \left[\frac{1}{2} + \frac{3}{2}\cos 2\omega\right] = 0,$$

$$\frac{d^2 \theta}{dt^2} + 2\frac{dl}{dt}\frac{d\theta}{dt} + n'^2 \left[\frac{3}{2}\sin 2\omega\right] = 0,$$

where

$$\omega = \theta - \theta'.$$

Now these equations are defective, for they have been formed by omitting certain terms from the complete equations as given in Lecture II. Hence, calling l_0, θ_0 the values of l, θ, which we have proved in Lecture IV to be solutions of the above equations, if we substitute l_0, θ_0 in the complete equations of Lecture II, residuals are left, say X and Y respectively. And if l, θ be solutions of the complete equations, and if we write

$$l = l_0 + \delta l,$$
$$\theta = \theta_0 + \delta\theta,$$

where δl, $\delta\theta$ are small quantities whose squares and products may be neglected in the first instance, we obtain the following equations for determining δl, $\delta\theta$, the corrections to approximate solutions l_0, θ_0 already found :—

$$X + \frac{d^2\delta l}{dt^2} + 2\frac{dl_0}{dt}\frac{d\delta l}{dt} - 2\frac{d\theta_0}{dt}\frac{d\delta\theta}{dt} - 3\frac{\mu}{a^3}e^{-3l_0}\delta l + 3n'^2\sin 2\omega\delta\theta = 0,$$

$$Y + \frac{d^2\delta\theta}{dt^2} + 2\frac{dl_0}{dt}\frac{d\delta\theta}{dt} + 2\frac{d\theta_0}{dt}\frac{d\delta l}{dt} \qquad\qquad + 3n'^2\cos 2\omega\delta\theta = 0.$$

Now let us write

$$\frac{d\theta_0}{dt} = 1 + n' + v, \qquad \frac{\mu}{r_0^3} = \frac{\mu}{a^3}e^{-3l_0} = c + w,$$

where
$$c = (1 + n')^2 + \frac{1}{2}n'^2.$$

The quantity v consists wholly of periodic terms of the form $\cos 2i\psi$ multiplied by small coefficients; w contains, besides periodic terms, a small constant term, which however might be removed if we were to choose c as the constant part of μ/r_0^3 in place of according to the definition above.

Let $\delta'l$, $\delta'\theta$ be quantities defined by the equations

$$X + \frac{d^2\delta'l}{dt^2} - 2(1 + n')\frac{d\delta'\theta}{dt} - 3c\delta l = 0,$$

$$Y + \frac{d^2\delta'\theta}{dt^2} + 2(1 + n')\frac{d\delta'l}{dt} \qquad = 0;$$

then $\delta'l$, $\delta'\theta$ are approximations to the complete corrections δl, $\delta\theta$, which if substituted in the equations that give those corrections will leave residuals, say X' and Y', where

$$X' = 2 \frac{dl_0}{dt} \frac{d\delta'l}{dt} - 2v \frac{d\delta'\theta}{dt} - 3w\delta'l + 3n'^2 \sin 2\omega\delta'\theta,$$

$$Y' = 2 \frac{dl_0}{dt} \frac{d\delta'\theta}{dt} + 2v \frac{d\delta'l}{dt} \qquad + 3n'^2 \cos 2\omega\delta'\theta.$$

We see that their value is known when $\delta'l$, $\delta'\theta$ are determined.

Now $\delta'l$, $\delta'\theta$ may be determined as follows.

Let $\qquad X = p_0 + \Sigma p_i \cos i\psi, \qquad Y = \Sigma q_i \sin i\psi,$

where i takes all positive integral values; and assume

$$\delta'l = a_0 + \Sigma a_i \cos i\psi, \qquad \delta'\theta = \Sigma b_i \sin i\psi.$$

Then substituting and equating coefficients, the constant term gives

$$p_0 - 3ca_0 = 0,$$

and the terms in $i\psi$ give

$$p_i - i^2 a_i - 2(1+n')\,ib_i - 3ca_i \qquad = 0,$$
$$q_i - i^2 b_i - 2(1+n')\,ia_i \qquad\qquad = 0;$$

the second of these may be written

$$2(1+n')\frac{q_i}{i} - 4(1+n')^2 a_i - 2(1+n')\,ib_i = 0;$$

subtract from the first and we have

$$p_i - 2(1+n')\frac{q_i}{i} - a_i\left[i^2 - 4(1+n')^2 + 3c\right] = 0,$$

or $\qquad\qquad a_i = \dfrac{p_i - 2(1+n')\dfrac{q_i}{i}}{i^2 - (1+n')^2 + \dfrac{3}{2}n'^2} \,;$

and $\qquad\qquad b_i = -\,2(1+n')\dfrac{a_i}{i} + \dfrac{q_i}{i^2}.$

We see that a_i, b_i will be of the same order of small quantities as p_i, q_i, in general. And therefore the coefficients of the terms of X', Y' will be of order higher than those of X, Y. Proceed then to determine further corrections $\delta''l$, $\delta''\theta$ satisfying the equations

$$X' + \frac{d^2\delta''l}{dt^2} - 2(1+n')\frac{d\delta''\theta}{dt} - 3c\delta''l = 0,$$

$$Y' + \frac{d^2\delta''\theta}{dt^2} + 2(1+n')\frac{d\delta''l}{dt} \qquad = 0;$$

then if $\delta'l + \delta''l$, $\delta'\theta + \delta''\theta$ are substituted in the complete equations for δl, $\delta\theta$ the residuals become

$$X'' = 2\frac{dl_0}{dt}\frac{d\delta''l}{dt} - 2v\frac{d\delta''\theta}{dt} - 3w\delta''l + 3n'^2\sin 2\omega\delta''\theta,$$

$$Y'' = 2\frac{dl_0}{dt}\frac{d\delta''\theta}{dt} + 2v\frac{d\delta''l}{dt} \qquad + 3n'^2\cos 2\omega\delta''\theta,$$

expressions which, if developed in series of cosines and sines of multiples of ψ, will have coefficients of higher order than the corresponding coefficients in X', Y'. The like process may be repeated until the residuals become insensible; we then have sensibly correct values of δl, $\delta\theta$, giving

$$l = l_0 + \delta l = l_0 + \delta'l + \delta''l + \ldots\ldots,$$
$$\theta = \theta_0 + \delta\theta = \theta_0 + \delta'\theta + \delta''\theta + \ldots\ldots$$

We may now take into account squares and products of the small quantities δl, $\delta\theta$ by treating $l_0 + \delta l$, $\theta_0 + \delta\theta$ as given approximate solutions just as we have here treated l_0, θ_0; substitute them in the complete equations of motion, and determine the residuals X, Y which they leave. These residuals will form the basis of a second approximation, and the operation may be repeated until no further correction is necessary. It is to be observed that if δl, $\delta\theta$ depend upon some such constant as the eccentricity of the Earth's orbit around the Sun, or the parallax of the Sun, then successive approximations yield correctly and separately the terms which depend upon the first, second, ... powers of that constant.

LECTURE VIII.

THE PARALLACTIC INEQUALITY.

WE shall now apply the method of the last lecture to find the terms in the Moon's coordinates which depend upon the parallax of the Sun.

The values of l, θ found in Lecture IV are

$$l_0 = \log_e (r/a) = -a_2 \cos 2\psi,$$
$$\theta_0 = nt + \epsilon + b_2 \sin 2\psi,$$

and these satisfy the equations of motion in which the terms involving the Sun's parallax are omitted. Hence the residuals they leave from the complete equations are

$$X = -\lambda n'^2 \frac{r}{a} \left\{ \frac{9}{8} \cos (\theta_0 - \theta') + \frac{15}{8} \cos 3 (\theta_0 - \theta') \right\},$$

$$Y = \lambda n'^2 \frac{r}{a} \left\{ \frac{3}{8} \sin (\theta_0 - \theta') + \frac{15}{8} \sin 3 (\theta_0 - \theta') \right\},$$

where

$$\lambda = \frac{E - M}{E + M} \frac{a}{a'}.$$

Now from above

$$\theta_0 - \theta' = \psi + b_2 \sin 2\psi\,;$$

hence we have

$$\sin (\theta_0 - \theta') = \sin \psi + \frac{1}{2} b_2 (\sin \psi + \sin 3\psi),$$

$$\cos (\theta_0 - \theta') = \cos \psi - \frac{1}{2} b_2 (\cos \psi - \cos 3\psi),$$

$$\sin 3 (\theta_0 - \theta') = \sin 3\psi + \frac{3}{2} b_2 (\sin \psi + \sin 5\psi),$$

$$\cos 3 (\theta_0 - \theta') = \cos 3\psi - \frac{3}{2} b_2 (\cos \psi - \cos 5\psi);$$

and

$$\frac{r}{a}\left\{\frac{9}{8}\cos(\theta_0-\theta')+\frac{15}{8}\cos 3(\theta_0-\theta')\right\}=\left(\frac{9}{8}-\frac{27}{8}b_2-\frac{3}{2}a_2\right)\cos\psi$$

$$+\left(\frac{15}{8}+\frac{9}{16}b_2-\frac{9}{16}a_2\right)\cos 3\psi+\left(\frac{45}{16}b_2-\frac{15}{16}a_2\right)\cos 5\psi,$$

$$\frac{r}{a}\left\{\frac{3}{8}\sin(\theta_0-\theta')+\frac{15}{8}\sin 3(\theta_0-\theta')\right\}=\left(\frac{3}{8}-\frac{21}{8}b_2-\frac{3}{4}a_2\right)\sin\psi$$

$$+\left(\frac{15}{8}+\frac{3}{16}b_2-\frac{3}{16}a_2\right)\sin 3\psi+\left(\frac{45}{16}b_2+\frac{15}{16}a_2\right)\sin 5\psi.$$

Assume

$$-\delta\lambda=\lambda a_1\cos\psi+\lambda a_3\cos 3\psi,$$
$$\delta\theta=\lambda b_1\sin\psi+\lambda b_3\sin 3\psi,$$

neglecting for the present the terms in 5ψ. In the present case it happens that it is more advantageous to substitute these expressions directly in the complete equations for δl, $\delta\theta$ given in the last lecture than to follow exactly the process for finding them by successive approximation. Omitting the factor λ, we get

$$a_1\cos\psi+9a_3\cos 3\psi+4a_2\sin 2\psi\,[a_1\sin\psi+3a_3\sin 3\psi]$$

$$+3\frac{\mu}{a^3}[a_1\cos\psi+a_3\cos 3\psi]+3\frac{\mu}{a^3}[3a_2\cos 2\psi][a_1\cos\psi+a_3\cos 3\psi]$$

$$-[2(1+n')+4b_2\cos 2\psi][b_1\cos\psi+3b_3\cos 3\psi]$$

$$+3n'^2[\sin 2\psi+b_2\sin 4\psi][b_1\sin\psi+b_3\sin 3\psi]$$

$$-n'^2\left(\frac{9}{8}-\frac{27}{8}b_2-\frac{3}{2}a_2\right)\cos\psi-n'^2\left(\frac{15}{8}+\frac{9}{16}b_2-\frac{9}{16}a_2\right)\cos 3\psi$$

$$-n'^2\left(\frac{45}{16}b_2-\frac{15}{16}a_2\right)\cos 5\psi=0,$$

$$-b_1\sin\psi-9b_3\sin 3\psi+4a_2\sin 2\psi\,[b_1\cos\psi+3b_3\cos 3\psi]$$

$$+3n'^2[-b_2+\cos 2\psi+b_2\cos 4\psi][b_1\sin\psi+b_3\sin 3\psi]$$

$$+[2(1+n')+4b_2\cos 2\psi][a_1\sin\psi+3a_3\sin 3\psi]$$

$$+n'^2\left(\frac{3}{8}-\frac{21}{8}b_2-\frac{3}{4}a_2\right)\sin\psi+n'^2\left(\frac{15}{8}+\frac{3}{16}b_2-\frac{3}{16}a_2\right)\sin 3\psi$$

$$+n'^2\left(\frac{45}{16}b_2-\frac{15}{16}a_2\right)\sin 5\psi=0.$$

If we equate to zero the coefficients of $\cos \psi$ and $\cos 3\psi$ in the first, and those of $\sin \psi$ and $\sin 3\psi$ in the second, we obtain the following equations for a_1, b_1, a_3, b_3; the terms in 5ψ remain outstanding, and the effect of a_5, b_5 in modifying the other coefficient is neglected.

$$a_1 \left[1 + 2a_2 + 3\frac{\mu}{a^3}\left(1 + \frac{3}{2}a_2\right) \right] - b_1 \left[2(1+n') + 2b_2 - \frac{3}{2}n'^2 \right]$$

$$+ a_3 \left[6a_2 + \frac{9}{2}\frac{\mu}{a^3}a_2 \right] - b_3 \left[6b_2 - \frac{3}{2}n'^2(1+b_2) \right] = \frac{3}{8}n'^2(3 - 9b_2 - 4a_2),$$

$$a_1 \left[2(1+n') - 2b_2 \right] - b_1 \left[1 + \frac{3}{2}n'^2 - 2a_2 + 3n'^2 b_2 \right]$$

$$+ a_3 \left[6b_2 \right] \qquad - b_3 \left[6a_2 - \frac{3}{2}n'^2 + \frac{3}{2}n'^2 b_2 \right] = -\frac{3}{8}n'^2(1 - 7b_2 - 2a_2),$$

$$a_1 \left[-2a_2 + \frac{9}{2}\frac{\mu}{a^3}a_2 \right] + b_1 \left[-2b_2 - \frac{3}{2}n'^2 + \frac{3}{2}n'^2 b_2 \right]$$

$$+ a_3 \left[9 + 3\frac{\mu}{a^3} \right] \qquad - b_3 [6(1+n')] \qquad = \frac{3}{8}n'^2\left(5 + \frac{3}{2}b_2 - \frac{3}{2}a_2\right),$$

$$a_1 [2b_2] + b_1 \left[2a_2 + \frac{3}{2}n'^2 - \frac{3}{2}n'^2 b_2 \right]$$

$$+ a_3 [6(1+n')] \qquad - b_3 [9 + 3n'^2 b_2] \qquad = -\frac{3}{8}n'^2\left(5 + \frac{1}{2}b_2 - \frac{1}{2}a_2\right),$$

If we require the formal values of a_1, b_1, a_3, b_3, we must substitute for $a_2, b_2, \mu/a^3$ the expressions we have found for them, and it will then be best to develope the coefficients in ascending powers of n'. But it is difficult to obtain by this process such good numerical results as we can get by substituting the numerical values of $a_2, b_2, \mu/a^3$ immediately in the equations above. If we do so we get the equations

$$4{\cdot}56672a_1 - 2{\cdot}17232b_1 + \quad{\cdot}08093a_3 - \quad{\cdot}05137b_3 = \frac{3}{8}n'^2 \times 2{\cdot}87937,$$

$$2{\cdot}14128a_1 - 0{\cdot}99564b_1 + \quad{\cdot}06127a_3 - \quad{\cdot}03338b_3 = -\frac{3}{8}n'^2 \times 0{\cdot}91416,$$

$${\cdot}02349a_1 - \quad{\cdot}03012b_1 + 12{\cdot}51451a_3 - 6{\cdot}48508b_3 = \frac{3}{8}n'^2 \times 5{\cdot}00455,$$

$${\cdot}02042a_1 + \quad{\cdot}02406b_1 + 6{\cdot}48508a_3 - 9{\cdot}00020b_3 = -\frac{3}{8}n'^2 \times 5{\cdot}00152.$$

We notice that the first equation is not very different from the second doubled; it is this fact that makes successive approximation a disadvantageous method and renders it advisable to include small quantities from the beginning.

Eliminate a_3, b_3 in succession from the third and fourth equations, thus:—

Multiply the third equation by

$$9{\cdot}00020 \div [12{\cdot}51451 \times 9{\cdot}00020 \quad 6{\cdot}48508 \times 6{\cdot}48508] = \quad 0{\cdot}127523,$$

and the fourth by

$$-6{\cdot}48508 \div [12{\cdot}51451 \times 9{\cdot}00020 - 6{\cdot}48508 \times 6{\cdot}48508] = -0{\cdot}091887,$$

and add; b_3 will be eliminated.

Again multiply the third equation by ${\cdot}091887$ and the fourth by $-{\cdot}177317$ and add; a_3 will be eliminated. Hence we find

$$ {\cdot}001119a_1 - {\cdot}006052b_1 + a_3 = \frac{3}{8}n'^2 \times 1{\cdot}09776, $$

$$ -{\cdot}001463a_1 - {\cdot}007034b_1 + b_3 = \frac{3}{8}n'^2 \times 1{\cdot}34669. $$

Multiply these by $-{\cdot}08093$ and ${\cdot}05137$ respectively and add to the first equation:—

$$ 4{\cdot}56672a_1 - 2{\cdot}17232b_1 + {\cdot}08093a_3 - {\cdot}05137b_3 = \frac{3}{8}n'^2 \times \quad 2{\cdot}87937, $$

$$ -0{\cdot}00009a_1 + 0{\cdot}00049b_1 - {\cdot}08093a_3 \qquad = \frac{3}{8}n'^2 \times -0{\cdot}08884, $$

$$ -0{\cdot}00008a_1 - 0{\cdot}00036b_1 \qquad + {\cdot}05137b_3 = \frac{3}{8}n'^2 \times \quad 0{\cdot}06918 ; $$

hence

$$ 4{\cdot}56665a_1 - 2{\cdot}17219b_1 = \frac{3}{8}n'^2 \times 2{\cdot}85971. $$

Eliminate a_3, b_3 in a similar manner from the second equation;

$$ 2{\cdot}14128a_1 - 0{\cdot}99546b_1 + {\cdot}06127a_3 - {\cdot}03338b_3 = -\frac{3}{8}n'^2 \times \quad 0{\cdot}91416, $$

$$ -0{\cdot}00007a_1 + 0{\cdot}00037b_1 - {\cdot}06127a_3 \qquad = -\frac{3}{8}n'^2 \times \quad 0{\cdot}06726, $$

$$ -0{\cdot}00005a_1 - 0{\cdot}00023b_1 \qquad + {\cdot}03338b_3 = -\frac{3}{8}n'^2 \times -0{\cdot}04495 ; $$

hence

$$2\!\cdot\!14116a_1 - 0\!\cdot\!99550b_1 = -\frac{3}{8}n'^2 \times 0\!\cdot\!93647.$$

From these equations we find

$$a_1 = -\frac{3}{8}n'^2 \times 46\!\cdot\!4814 \;\; = -\;\cdot\!11392,8,$$

$$b_1 = -\frac{3}{8}n'^2 \times 99\!\cdot\!0336 \;\; = -\;\cdot\!24273,4,$$

$$a_3 = \;\;\frac{3}{8}n'^2 \times \;\;\cdot\!55042 = \;\;\cdot\!00134,9,$$

$$b_3 = \;\;\frac{3}{8}n'^2 \times \;\;\cdot\!58209 = \;\;\cdot\!00142,7.$$

LECTURE IX.

THE PARALLACTIC INEQUALITY (*continued*).

LET us now consider the terms in 5ψ which have been left outstanding.

Include additional terms $\lambda a_5 \cos 5\psi$, $\lambda b_5 \sin 5\psi$ in $-\delta l$, $\delta\theta$, and equate to zero the coefficients of $\cos 5\psi$, $\sin 5\psi$ in the differential equations that give δl, $\delta\theta$.

We have

$$25a_5 + \frac{3\mu}{a^3} a_5 - 10(1 + n')b_5 - 6a_2a_3 - 6b_2b_3 + \frac{9}{2}\frac{\mu}{a^3} a_2a_3$$
$$- \frac{3}{2} n'^2 b_3 - n'^2 \left(\frac{45}{16} b_2 - \frac{15}{16} a_2\right) = 0,$$

$$-25b_5 \qquad + 10(1 + n')a_5 + 6a_2b_3 + 6a_3b_2$$
$$+ \frac{3}{2} n'^2 b_3 + n'^2 \left(\frac{45}{16} b_2 - \frac{15}{16} a_2\right) = 0.$$

In these equations substitute

$$\frac{\mu}{a^3} = 1 + 2n' + \frac{3}{2} n'^2.$$

Then

$$\left(28 + 6n' + \frac{9}{2} n'^2\right) a_5 - 10(1 + n')b_5 = \left(\frac{3}{2} - 9n' - \frac{27}{4} n'^2\right) a_2a_3$$
$$+ \left(6b_2 + \frac{3}{2} n'^2\right) b_3 + n'^2 \left(\frac{45}{16} b_2 - \frac{15}{16} a_2\right),$$

$$-25b_5 + 10(1 + n')a_5 = -6a_2b_3 - 6b_2a_3 - \frac{3}{2} n'^2 b_3 - n'^2 \left(\frac{45}{16} b_2 - \frac{15}{16} a_2\right).$$

Eliminate b_5:

$$\left(24 - 2n' + \frac{1}{2} n'^2\right) a_5 = \left[\left(\frac{3}{2} - 9n' - \frac{27}{4} n'^2\right) a_2 + \frac{12}{5} (1 + n') b_2\right] a_3$$

$$+ \left[6b_2 + \frac{3}{2} n'^2 + \frac{12}{5} (1 + n') a_2 + \frac{3}{5} (1 + n') n'^2\right] b_3$$

$$+ (7 + 2n') n'^2 \left(\frac{9}{16} b_2 - \frac{3}{16} a_2\right),$$

and then b_5 is given by

$$b_5 = \frac{2}{5} (1 + n') a_5 + \frac{6}{25} b_2 a_3 + \left(\frac{6}{25} a_2 + \frac{3}{50} n'^2\right) b_3 + \frac{1}{5} n'^2 \left(\frac{9}{16} b_2 - \frac{3}{16} a_2\right).$$

From these we find

$$a_5 = \frac{3}{8} n'^2 \times \cdot 00595,3 = \cdot 00001,4591,$$

$$b_5 = \frac{3}{8} n'^2 \times \cdot 00710,3 = \cdot 00001,7410.$$

These numbers being so small, we see that we may safely ignore, as we have done, their effect in modifying the earlier coefficients.

To find the effect of these coefficients upon the Moon's coordinates we must multiply by the factor $\lambda = \dfrac{E - M}{E + M} \cdot \dfrac{a}{a'}$.

We shall take in accordance with the results given in *Monthly Notices*, Vol. 13, p. 177, and Appendix to the Nautical Almanac, 1856,

$$\frac{E}{M} = 81\cdot 5.$$

Constant of Moon's Parallax $= 3422''\cdot 325$.

Also we shall take in the first place, the Sun's Mean Parallax to be $8'''\cdot 8$, and in the next place $8'''\cdot 9$, and we will find the corresponding values of the coefficients of the Parallactic Inequalities.

We find

8″·8		8″·9

$\lambda = \;$ ·00250,9 $\lambda = \;$ ·00253,76

$\lambda a_1 = - \;$ ·00028,585 $\lambda a_1 = - \;$ ·00028,910

$\lambda b_1 = - \;$ ·00060,903 $= -125''\!\cdot\!62$ $\lambda b_1 = - \;$ ·00061,596

$= -127''\!\cdot\!05$

$\lambda a_3 = \;$ ·00000,3385 $\lambda a_3 = \;$ ·00000,3423

$\lambda b_3 = \;$ ·00000,3580 $=$ $0''\!\cdot\!7384$ $\lambda b_3 = \;$ ·00000,3620

$=$ $0''\!\cdot\!7468$

$\lambda a_5 = \;$ ·00000,00366 $\lambda a_5 = \;$ ·00000,00370

$\lambda b_5 = \;$ ·00000,00437 $=$ $0''\!\cdot\!00901$ $\lambda b_5 = \;$ ·00000,00442

$=$ $0''\!\cdot\!00911.$

These results are very fairly accurate; but in order to get good values for a_1, b_1, we were obliged to discuss a_1, b_1, a_3, b_3 simultaneously. Let us consider the peculiarity of the equations from which this difficulty arose.

Following the method of approximation of Lecture VII, if we neglect at first the products of δl, $\delta \theta$, $d\delta l/dt$, $d\delta \theta/dt$ with the small quantities a_2, b_2, n'^2, the equations become

$$\frac{d^2\delta l}{dt^2} - 2n\frac{d\delta\theta}{dt} - 3\frac{\mu}{a^3}\delta l + X = 0,$$

$$\frac{d^2\delta\theta}{dt^2} + 2n\frac{d\delta l}{dt} \qquad\quad + Y = 0.$$

Now suppose the following is a set of terms that appear

in X $p_i \cos(it + \gamma)$,   in Y $q_i \sin(it + \gamma)$,

δl $a_i \cos(it + \gamma)$, $\delta \theta$ $b_i \sin(it + \gamma)$;

then as in Lecture VII, we find

$$a_i = \frac{p_i - 2\dfrac{n}{i}q_i}{i^2 - n^2 + \dfrac{3}{2}n'^2}.$$

$$b_i = -2\frac{n}{i}a_i + \frac{1}{i^2}q.$$

Therefore if i differs little from n, the divisor in a_i will be small, and a small error or omission in the numerator of a_i will appear magnified in the values of both a_i and b_i. In the case of the first term of the Parallactic Inequality,

$$i = n - n',$$

$$i^2 - n^2 + \frac{3}{2} n'^2 = -2nn' + \frac{5}{2} n'^2;$$

and if we take

$$p_i = -\frac{9}{8} n'^2, \quad q_i = \frac{3}{8} n'^2,$$

which differ from the correct values by quantities of the fourth order, then

$$p_i - 2\frac{n}{i} q_i = -\frac{3}{8} n'^2 \frac{5n - 3n'}{n - n'},$$

and the formulae give

$$a_i = \frac{3}{4} \frac{n'(5n - 3n')}{(n - n')(4n - 5n')},$$

$$b_i = -\frac{2n}{n - n'} a_i + \frac{3}{8} \frac{n'^2}{(n - n')^2}.$$

Now if we develope these expressions in ascending powers of m, i.e. n'/n, the first terms are

$$a_i = \frac{15}{16} m, \quad b_i = -\frac{15}{8} m,$$

and these are the only terms which the formulae derived from our method of approximation will give correctly.

LECTURE X.

THE ANNUAL EQUATION.

LET us next take into account the effect of the first power of the eccentricity of the Earth's orbit. We shall find that it produces an inequality in the Moon's coordinates, the chief part of which has a period of one year, and is therefore called the Annual Equation.

In the formulae of Lecture VII, let the known approximate solutions l_0, θ_0, include the Variation only; then the equations for the corrections δl, $\delta \theta$ are

$$X + \frac{d^2 \delta l}{dt^2} + 4a_2 \sin 2\psi \frac{d\delta l}{dt} - 3\frac{\mu}{a^3}(1 + 3a_2 \cos 2\psi)\, \delta l$$

$$- 2\left[(1 + n') + 2b_2 \cos 2\psi\right] \frac{d\delta\theta}{dt} + 3n'^2(\sin 2\psi + b_2 \sin 4\psi)\delta\theta = 0,$$

$$Y + \frac{d^2 \delta\theta}{dt^2} + 4a_2 \sin 2\psi \frac{d\delta\theta}{dt} + 3n'^2(-b_2 + \cos 2\psi + b_2 \cos 4\psi)\, \delta\theta$$

$$+ 2\left[(1 + n') + 2b_2 \cos 2\psi\right] \frac{d\delta l}{dt} \qquad\qquad = 0,$$

where a_2, b_2, μ/a^3 are known quantities whose values are given in Lectures IV, V.

Refer now to Lecture III, and we find that the terms that are left outstanding when the terms of the Variation are substituted, and the parallactic terms omitted are the following:—

$$X = -\frac{3}{2} n'^2 e' \cos(n't - \varpi') - \frac{21}{4} n'^2 e' \cos\{2(\theta - n't) - (n't - \varpi')\}$$

$$+ \frac{3}{4} n'^2 e' \cos\{2(\theta - n't) + (n't - \varpi')\}$$

$$Y = \qquad\qquad + \frac{21}{4} n'^2 e' \sin\{2(\theta - n't) - (n't - \varpi')\}$$

$$- \frac{3}{4} n'^2 e' \sin\{2(\theta - n't) + (n't - \varpi')\}.$$

Write α for $n't - \varpi'$; then

$$\cos\{2(\theta - n't) - \alpha\} = \quad \cos(2\psi - \alpha) - (2b_2 \sin 2\psi) \sin(2\psi - \alpha)$$

$$= -b_2 \cos\alpha + \cos(2\psi - \alpha) + b_2 \cos(4\psi - \alpha),$$

$$\sin\{2(\theta - n't) - \alpha\} = +b_2 \sin\alpha + \sin(2\psi - \alpha) + b_2 \sin(4\psi - \alpha),$$

$$\cos\{2(\theta - n't) + \alpha\} = -b_2 \cos\alpha + \cos(2\psi + \alpha) + b_2 \cos(4\psi + \alpha),$$

$$\sin\{2(\theta - n't) + \alpha\} = -b_2 \sin\alpha + \sin(2\psi + \alpha) + b_2 \sin(4\psi + \alpha).$$

Hence

$$X = -\frac{3}{2} n'^2 (1 - 3b_2) e' \cos\alpha - \frac{21}{4} n'^2 e' \cos(2\psi - \alpha)$$

$$+ \frac{3}{4} n'^2 e' \cos(2\psi + \alpha) - \frac{21}{4} n'^2 b_2 e' \cos(4\psi - \alpha) + \frac{3}{4} n'^2 b_2 e' \cos(4\psi + \alpha),$$

$$Y = 6n'^2 b_2 e' \sin\alpha + \frac{21}{4} n'^2 e' \sin(2\psi - \alpha) - \frac{3}{4} n'^2 e' \sin(2\psi + \alpha)$$

$$+ \frac{21}{4} n'^2 b_2 e' \sin(4\psi - \alpha) - \frac{3}{4} n'^2 b_2 e' \sin(4\psi + \alpha).$$

For our present purpose we shall ignore the small terms in $4\psi - \alpha$ and $4\psi + \alpha$ which are of the sixth order.

Assume

$$-\delta l = a_5 e' \cos\alpha + a_6 e' \cos(2\psi - \alpha) + a_7 e' \cos(2\psi + \alpha),$$

$$\delta\theta = b_5 e' \sin\alpha + b_6 e' \sin(2\psi - \alpha) + b_7 e' \sin(2\psi + \alpha).$$

Now the terms which arise in the left-hand members of the

equations owing to terms $-a_p \cos pt$ in δl, and $b_p \sin pt$ in $\delta\theta$, will be

$$p^2 a_p \cos pt + 2a_2\, pa_p\left[\cos(pt - 2\psi) - \cos(pt + 2\psi)\right]$$

$$+\frac{3\mu}{a^3}\,a_p\left[\cos pt + \frac{3}{2}\,a_2\cos(pt - 2\psi) + \frac{3}{2}\,a_2\cos(pt + 2\psi)\right]$$

$$-2(1 + n')\,pb_p\cos pt - 2b_2\,pb_p\left[\cos(pt - 2\psi) + \cos(pt + 2\psi)\right]$$

$$+\frac{3}{2}\,n'^2 b_p\left[\cos(pt - 2\psi) - \cos(pt + 2\psi)\right],$$

and

$$-p^2 b_p\sin pt + 2a_2\,pb_p\left[-\sin(pt - 2\psi) + \sin(pt + 2\psi)\right]$$

$$+\frac{3}{2}\,n'^2 b_p\left[\sin(pt - 2\psi) - 2b_2\sin pt + \sin(pt + 2\psi)\right]$$

$$+2(1 + n')\,pa_p\sin pt + 2b_2\,pa_p\left[\sin(pt - 2\psi) + \sin(pt + 2\psi)\right],$$

respectively, neglecting the very small quantities in 4ψ.

Hence we get the equations following :—

Equate to zero the coefficients of $\cos\alpha$, $\sin\alpha$:

$$\left(n'^2 + \frac{3\mu}{a^3}\right)a_5 - 2n'(1 + n')\,b_5 + 2(2 - n')\,a_2 a_6 + \frac{3\mu}{a^3}\cdot\frac{3}{2}\,a_2 a_6$$

$$-2(2 - n')\,b_2 b_6 + \frac{3}{2}\,n'^2 b_6 + 2(2 + n')\,a_2 a_7 + \frac{3\mu}{a^3}\cdot\frac{3}{2}\,a_2 a_7$$

$$-2(2 + n')\,b_2 b_7 + \frac{3}{2}\,n'^2 b_7 = \frac{3}{2}\,n'^2(1 - 3b_2)$$

$$-n'^2 b_5 - 3n'^2 b_2 b_5 + 2(1 + n')\,n'a_5 + 2(2 - n')\,a_2 b_6$$

$$-2(2 - n')\,b_2 a_6 - \frac{3}{2}\,n'^2 b_6 - 2(2 + n')\,a_2 b_7$$

$$+2(2 + n')\,b_2 a_7 + \frac{3}{2}\,n'^2 b_7 = -6n'^2 b_2.$$

Equate the coefficients of $\cos(2\psi - \alpha)$, $\sin(2\psi - \alpha)$:—

$$(2 - n')^2 a_6 + 3\frac{\mu}{a^3}\,a_6 - 2(1 + n')(2 - n')\,b_6 + 2n'a_2 a_5 + 3\frac{\mu}{a^3}\cdot\frac{3}{2}\,a_2 a_5$$

$$-2n'b_2 b_5 + \frac{3}{2}\,n'^2 b_5 = \frac{21}{4}\,n'^2,$$

$$-(2 - n')^2 b_6 - 3n'^2 b_2 b_6 + 2(1 + n')(2 - n')\,a_6 + 2n'a_2 b_5 - 2n'b_2 a_5$$

$$-\frac{3}{2}\,n'^2 b_5 = -\frac{21}{4}\,n'^2.$$

Equate the coefficients of $\cos(2\psi + \alpha)$, $\sin(2\psi + \alpha)$:—

$$(2+n')^2 a_7 + 3\frac{\mu}{a^3} a_7 - 2(1+n')(2+n') b_7 - 2n' a_2 a_5 + 3\frac{\mu}{a^3} \cdot \frac{3}{2} a_2 a_5$$

$$- 2n' b_2 b_5 - \frac{3}{2} n'^2 b_5 = -\frac{3}{4} n'^2,$$

$$-(2+n')^2 b_7 - 3n'^2 b_2 b_7 + 2(1+n')(2+n') a_7 + 2n' a_2 b_5 + 2n' b_2 a_5$$

$$+ \frac{3}{2} n'^2 b_5 = \frac{3}{4} n'^2.$$

In equations of this class, as a general rule we would determine a_5, b_5 approximately from the first pair, substitute them in the second pair and determine a_6, b_6 approximately, and similarly a_7, b_7 from the third pair, and repeat this approximation as often as might be necessary. But if we refer to the second equation, we see that b_5 must be determined by means of a small divisor, and this puts any method of approximation at a disadvantage. In order to obtain readily satisfactory values for the new coefficients, we shall treat the six equations simultaneously, substituting first the numerical values of the known quantities.

We have found

$$a_2 = \cdot00717,95, \quad b_2 = \cdot01021,20, \quad \mu/a^3 = 1\cdot17150,3.$$

Hence

$$3\cdot52105a_5 - 0\cdot174763b_5 + \cdot065405a_6 - \cdot029393b_6 + \cdot067727a_7$$
$$- \cdot032695b_7 = \frac{3}{2} n'^2 \times \quad 0\cdot969364,$$

$$0\cdot174763a_5 - 0\cdot006736b_5 - \cdot039197a_6 + \cdot017753b_6 + \cdot042499a_7$$
$$- \cdot020075b_7 = \frac{3}{2} n'^2 \times -0\cdot040848,$$

$$\cdot039009a_5 + \quad\cdot008153b_5 + 7\cdot19766a_6 - 4\cdot14858b_6$$
$$= \frac{3}{2} n'^2 \times \quad 3\cdot50,$$

$$- \quad\cdot001651a_5 - \quad\cdot008643b_5 + 4\cdot14858a_6 - 3\cdot68335b_6$$
$$= \frac{3}{2} n'^2 \times -3\cdot50,$$

$$\cdot036687a_5 - \cdot011455b_5 \qquad\qquad + 7\cdot84443a_7 - 4\cdot49811b_7$$
$$= \frac{3}{2}n'^2 \times - 0\cdot50,$$

$$\cdot001651a_5 + \cdot010965b_5 \qquad\qquad + 4\cdot49811a_7 - 4\cdot33012b_7$$
$$= \frac{3}{2}n'^2 + \quad 0\cdot50.$$

From the second and third pairs we find

$$\cdot016186a_5 + \cdot007084b_5 + a_6 = \frac{3}{2}n'^2 \times \quad 2\cdot94739,$$

$$\cdot018678a_5 + \cdot010326b_5 + b_6 = \frac{3}{2}n'^2 \times \quad 4\cdot26994,$$

$$\cdot011026a_5 - \cdot007203b_5 + a_7 = \frac{3}{2}n'^2 \times - 0\cdot321398,$$

$$\cdot011072a_5 - \cdot010015b_5 + b_7 = \frac{3}{2}n'^2 \times - 0\cdot449338.$$

Eliminate a_6, b_6, a_7, b_7 from the first pair.

Hence

$$3\cdot52015a_5 - \cdot174762b_5 = \frac{3}{2}n'^2 \times 0\cdot909170,$$

$$0\cdot174819a_5 - \cdot006536b_5 = \frac{3}{2}n'^2 \times 0\cdot003516.$$

Hence

$$a_5 = \frac{3}{2}n'^2 \times - \quad 0\cdot70619,8 = - \cdot00692,37,$$

$$b_5 = \frac{3}{2}n'^2 \times - 19\cdot4268 \quad\;\; = - \cdot19046,3.$$

Now e' is a constant found by observation; taking $e' = 3459''\cdot28$, its value in 1850, we get

$$a_5 e' = - \cdot00011,61,$$
$$b_5 e' = - \cdot00319,4 \quad = - 658''\cdot9,$$

and further

$$a_6 = \quad \cdot03035,8, \quad a_6 e' = \quad \cdot00050,9,$$
$$b_6 = \quad \cdot04396,7, \quad b_6 e' = \quad \cdot00073,73 = \quad 152''\cdot09,$$
$$a_7 = - \cdot00444,7, \quad a_7 e' = - \cdot00007,457,$$
$$b_7 = - \cdot00623,6, \quad b_7 e' = - \cdot00010,46 = - \quad 21'''\cdot57.$$

LECTURE XI.

THE EQUATION OF THE CENTRE AND THE EVECTION.

We have seen that the equations of motion

$$\frac{d^2l}{dt^2} + \left(\frac{dl}{dt}\right)^2 - \left(\frac{d\theta}{dt}\right)^2 + \frac{\mu}{r^3} - n'^2\left[\frac{1}{2} + \frac{3}{2}\cos 2\left(\theta - n't\right)\right] = 0,$$

$$\frac{d^2\theta}{dt^2} + 2\frac{d\theta}{dt}\frac{dl}{dt} \qquad + n'^2\left[\qquad \frac{3}{2}\sin 2\left(\theta - n't\right)\right] = 0,$$

are satisfied very approximately by the values

$$l = \log\frac{r}{a} = -a_2\cos 2\psi,$$

$$\theta \qquad = \quad nt + \epsilon + b_2\sin 2\psi,$$

where $\qquad \psi = nt + \epsilon - (n't + \epsilon'),$

and a_2, b_2 are small quantities depending upon the ratio n'/n, and a is a quantity depending upon n in such a way that

$$\frac{\mu}{a^3} = n^2 + \frac{1}{2}n'^2 - \frac{9}{32}\frac{n'^4}{(n-n')^2} + 2n'(2n-n')\,a_2^2,$$

while n, ϵ are arbitrary, though subject to the assumption that the ratio n'/n is small.

This solution, then, expresses a possible case of motion; nevertheless it is no more than a particular case because it involves only two arbitrary constants, whereas the complete and general solution must contain four, in order that it may be able to satisfy any given initial conditions, that is, in order that the initial coordinates and their initial velocities may have any given values.

When there is no disturbance the four arbitrary constants are n and ϵ,—which denote quantities similar to those expressed by the same symbols above,—and the two elliptic elements e and ϖ, of which e denotes the eccentricity of the orbit and ϖ the longitude of the apse.

We will now shew how to complete the solution by introducing into $\log(a/r)$ and θ additional terms depending on quantities similar to e, ϖ, of which the former is constant and the latter varies slowly and uniformly with t; and for the sake of simplicity we will suppose at first that e is so small that its square and higher powers may be neglected though it is otherwise arbitrary in magnitude.

Let us assume then

$$\log \frac{a}{r} = \qquad a_2 \cos 2\psi + e \cos (nt - \varpi),$$

$$\theta = nt + \epsilon + b_2 \sin 2\psi + 2e(1 + b_0)\sin(nt - \varpi),$$

in which the elliptic terms are of the same form as in the undisturbed orbit, and ϖ is supposed to be slowly variable, so that

$$\frac{d\varpi}{dt} = p,$$

where p is supposed to be a small quantity of the order of the disturbing force.

We will now substitute these assumed values in the differential equations. We have

$$\frac{dl}{dt} = \qquad 2(n - n')a_2 \sin 2\psi + (n - p)e\sin(nt - \varpi),$$

$$\frac{d^2l}{dt^2} = \qquad 4(n - n')^2 a_2 \cos 2\psi + (n - p)^2 e \cos(nt - \varpi),$$

$$\frac{d\theta}{dt} = n + 2(n - n')b_2 \cos 2\psi + 2(n - p)(1 + b_0)e\cos(nt - \varpi),$$

$$\frac{d^2\theta}{dt^2} = \qquad -4(n - n')b_2 \sin 2\psi - 2(n - p)^2(1 + b_0)e\sin(nt - \varpi).$$

Hence

$$4(n - n')^2 a_2 \cos 2\psi + (n - p)^2 e\cos(nt - \varpi)$$
$$+ 2(n - n')(n - p)a_2 e\left[\cos(2\psi - nt + \varpi) - \cos(2\psi + nt - \varpi)\right]$$

$$- \{n^2 + 4n(n-n')\, b_2 \cos 2\psi + 4n(n-p)(1+b_0)\, e \cos (nt - \varpi)$$

$$+ 4(n-n')(n-p)(1+b_0) b_2 e [\cos (2\psi - nt + \varpi) + \cos(2\psi + nt - \varpi)]\}$$

$$+ \frac{\mu}{a^3} \{1 + 3a_2 \cos 2\psi\} \{1 + 3e \cos (nt - \varpi)\}$$

$$- n'^2 \left\{ \frac{1}{2} + \frac{3}{2} \cos 2\psi - 3 \sin 2\psi \, [2(1+b_0)\, e \sin (nt - \varpi)] \right\} = 0,$$

and

$$-4(n-n')^2\, b_2 \sin 2\psi - 2(n-p)^2 (1+b_0)\, e \sin (nt - \varpi)$$

$$+ 4n(n-n')\, a_2 \sin 2\psi + 2n(n-p)\, e \sin (nt - \varpi)$$

$$+ 4(n-n')(n-p)(1+b_0)\, ea_2 [\sin (2\psi - nt + \varpi) + \sin(2\psi + nt - \varpi)]$$

$$+ 2(n-n')(n-p)\, eb_2 [-\sin (2\psi - nt + \varpi) + \sin (2\psi + nt - \varpi)]$$

$$+ n'^2 \left\{ \frac{3}{2} \sin 2\psi + 3 \cos 2\psi \, [2(1+b_0)\, e \sin (nt - \varpi)] \right\} = 0.$$

It will of course be found that with the values of a_2, b_2 of Lecture IV, the terms independent of e vanish identically.

Equating to zero the coefficients of $\cos (nt - \varpi)$ in the first equation and $\sin (nt - \varpi)$ in the second, we get

$$(n-p)^2 - 4n(n-p)(1+b_0) + \frac{3\mu}{a^3} = 0,$$

$$-2(n-p)^2 (1+b_0) + 2n(n-p) \quad\quad = 0.$$

Therefore $(n-p)(1+b_0) = n,$

and $(n-p)^2 = 4n^2 - \dfrac{3\mu}{a^3}$

$$= n^2 - \frac{3}{2}\, n'^2, \text{ approximately,}$$

or $\dfrac{p}{n} = \dfrac{3}{4}\, m^2 = b_0$, approximately.

Now terms have been left outstanding with the arguments $2\psi - nt + \varpi$, $2\psi + nt - \varpi$. These may be removed by assuming

$$\log \frac{a}{r} = a_2 \cos 2\psi + e \cos (nt - \varpi) + a_{21} e \cos (2\psi - nt + \varpi)$$

$$+ a_{22} e \cos (2\psi + nt - \varpi),$$

$$\theta = nt + \epsilon + b_2 \sin 2\psi + 2e(1+b_0) \sin (nt - \varpi)$$

$$+ b_{21} e \sin (2\psi - nt + \varpi) + b_{22} e \sin (2\psi + nt - \varpi).$$

Hence in place of the former equations, we get the following

$$(n-p)^2 - 4n(n-p)(1+b_0) + \frac{3\mu}{a^3}$$

$$+ \left[2(n-n')(n-2n'+p) + \frac{9}{2}\frac{\mu}{a^3} \right] a_2 a_{21}$$

$$+ \left[2(n-n')(3n-2n'-p) + \frac{9}{2}\frac{\mu}{a^3} \right] a_2 a_{22}$$

$$+ \left[-2(n-n')(n-2n'+p) b_2 + \frac{3}{2} n'^2 \right] b_{21}$$

$$+ \left[-2(n-n')(3n-2n'-p) b_2 + \frac{3}{2} n'^2 \right] b_{22} = 0,$$

$$-2(n-p)^2(1+b_0) + 2n(n-p) - 6n'^2 b_2(1+b_0)$$

$$- 2(n-n')(n-2n'+p) b_2 a_{21} + 2(n-n')(3n-2n'-p) b_2 a_{22}$$

$$+ \left[2(n-n')(n-2n'+p) a_2 - \frac{3}{2} n'^2 \right] b_{21}$$

$$+ \left[-2(n-n')(3n-2n'-p) a_2 + \frac{3}{2} n'^2 \right] b_{22} = 0.$$

Multiply the second by $-\dfrac{2n}{n-p}$, and add to the first; this will eliminate $1+b_0$.

$$(n-p)^2 - 4n^2 + \frac{3\mu}{a^3} + \frac{12nn'^2}{n-p} b_2(1+b_0)$$

$$+ 2(n-n')(n-2n'+p)[a_2 a_{21} - b_2 b_{21}]$$

$$+ \frac{9}{2}\frac{\mu}{a^3}[a_2 a_{21} + a_2 a_{22}] + \frac{3}{2} n'^2[b_{21} + b_{22}]$$

$$+ 2(n-n')(3n-2n'-p)[a_2 a_{22} - b_2 b_{22}]$$

$$+ 4\frac{n}{n-p}(n-n')(n-2n'+p)[b_2 a_{21} - a_2 b_{21}]$$

$$- 4\frac{n}{n-p}(n-n')(3n-2n'-p)[b_2 a_{22} - a_2 b_{22}]$$

$$+ 3n'^2\frac{n}{n-p}[b_{21} - b_{22}] = 0.$$

Also the equations obtained by equating the coefficients of $e \cos (2\psi - nt + \varpi)$ and $e \sin (2\psi - nt + \varpi)$ to zero are

$$\left[2 (n - n') (n - p) + \frac{9}{2} \frac{\mu}{a^3} \right] a_2 - 4 (n - n') (n - p) (1 + b_0) b_2 + 3n'^2 (1 + b_0)$$

$$+ \left[(n - 2n' + p)^2 + \frac{3\mu}{a^3} \right] a_{21} - 2n (n - 2n' + p) b_{21} = 0,$$

$$4 (n - n') (n - p) (1 + b_0) a_2 - 2 (n - n') (n - p) b_2$$
$$- 3n'^2 (1 + b_0) - (n - 2n' + p)^2 b_{21} + 2n (n - 2n' + p) a_{21} = 0.$$

Multiply the second by $- \dfrac{2n}{n - 2n' + p}$, and add to the first; this will eliminate b_{21}, and gives

$$\left[2 (n - n') (n - p) + \frac{9}{2} \frac{\mu}{a^3} - \frac{8n}{n - 2n' + p} (n - n') (n - p) (1 + b_0) \right] a_2$$

$$+ \left[- 4 (n - n') (n - p) (1 + b_0) + 4n \frac{n - n'}{n - 2n' + p} (n - p) \right] b_2$$

$$+ \left[3n'^2 + \frac{6nn'^2}{n - 2n' + p} \right] (1 + b_0) + \left[(n - 2n' + p)^2 - 4n^2 + \frac{3\mu}{a^3} \right] a_{21} = 0.$$

Lastly the equations obtained by equating the coefficients of
$$e \cos (2\psi + nt - \varpi) \text{ and } e \sin (2\psi + nt - \varpi)$$
to zero are

$$\left[- 2 (n - n') (n - p) + \frac{9}{2} \frac{\mu}{a^3} \right] a_2 - 4 (n - n') (n - p) (1 + b_0) b_2 - 3n'^2 (1 + b_0)$$

$$+ \left[(3n - 2n' - p)^2 + \frac{3\mu}{a^3} \right] a_{22} - 2n (3n - 2n' - p) b_{22} = 0,$$

and

$$4 (n - n') (n - p) (1 + b_0) a_2 + 2 (n - n') (n - p) b_2 + 3n'^2 (1 + b_0)$$
$$- (3n - 2n' - p)^2 b_{22} + 2n (3n - 2n' - p) a_{22} = 0.$$

Multiply the second by $- \dfrac{2n}{3n - 2n' - p}$ and add to the first; this will eliminate b_{22}, and gives

$$\left[-2(n-n')(n-p) + \frac{9}{2}\frac{\mu}{a^3} - \frac{8n}{3n-2n'-p}(n-n')(n-p)(1+b_0) \right] a_2$$

$$+ \left[-4(n-n')(n-p)(1+b_0) - \frac{4n}{3n-2n'-p}(n-n')(n-p) \right] b_2$$

$$+ \left[-3n'^2 - \frac{6nn'^2}{3n-2n'-p} \right](1+b_0)$$

$$+ \left[(3n-2n'-p)^2 - 4n^2 + \frac{3\mu}{a^3} \right] a_{22} = 0.$$

These six equations are to be solved by successive approximation; taking the first rough values of p/n and b_0, we find from the last two pairs values for a_{21}, b_{21}, a_{22}, b_{22}; these are substituted in the first pair and yield more approximate values of p/n and b_0, and so on.

It will be noticed that this complexity is made necessary by the fact that a_{21}, b_{21} are found by means of a small divisor

$$(n-2n'+p)^2 - 4n^2 + \frac{3\mu}{a^3}.$$

$$\left. \quad \right] \left[\quad \right]$$

LECTURE XII.

THE EVECTION AND THE MOTION OF THE APSE.

We proceed to the conversion into numbers of the formulae of Lecture XI.

Take
$$n - n' = 1,$$
$$n = 1\cdot08084,9,$$
$$n' = \cdot08084,9,$$
$$\frac{\mu}{a^3} = 1\cdot17150,3,$$
$$\log a_2 = 7\cdot85609,$$
$$\log b_2 = 8\cdot00911.$$

First Approximation.

$$(n - p)^2 - 4n^2 + 3\frac{\mu}{a^3} = 0,$$

$$
\begin{aligned}
4n^2 &\quad 4\cdot67293,7, \\
-3\frac{\mu}{a^3} &\quad -3\cdot51450,9, \\
(n - p)^2 = &\quad 1\cdot15842,8, \\
n - p = &\quad 1\cdot07630,3, \\
p = &\quad \cdot00454,6, \\
n - 2n' + p = &\quad \cdot92369,7, \\
3n - 2n' - p = &\quad 3\cdot07630,3, \\
1 + b_0 = &\quad 1\cdot00422,4 = n/(n - p).
\end{aligned}
$$

Substitute in the equation for a_{21},

$$\left[(n - 2n' + p)^2 - 4n^2 + \frac{3\mu}{a^3}\right] a_{21} + 2(n - n')(n - p) a_2 + \frac{9}{2}\frac{\mu}{a^3} a_2$$

$$-8\frac{n}{n - 2n' + p}(n - n')(n - p)(1 + b_0) a_2 - 4(n - n')(n - p)(1 + b_0) b_2$$

$$+4\frac{n}{n - 2n' + p}(n - n')(n - p) b_2 + 3n'^2(1 + b_0)$$

$$+6\frac{nn'^2}{n - 2n' + p}(1 + b_0) = 0.$$

The various terms give

$$2(n - n')(n - p) a_2 \qquad \qquad \cdot 01545,45$$

$$\frac{9}{2}\frac{\mu}{a^3} a_2 \qquad \qquad \cdot 03784,83$$

$$-8\frac{n}{n - 2n' + p}(n - n')(n - p)(1 + b_0) a_2 \qquad -\cdot 07264,10$$

$$-4(n - n')(n - p)(1 + b_0) b_2 \qquad -\cdot 04415,05$$

$$4\frac{n}{n - 2n' + p}(n - n')(n - p) b_2 \qquad \cdot 05144,47$$

$$3n'^2(1 + b_0) \qquad \qquad \cdot 01969,25$$

$$6\frac{nn'^2}{n - 2n' + p}(1 + b_0) \qquad \cdot 04608,58$$

$$\overline{\qquad \cdot 05373,43}$$

$$(n - 2n' + p)^2 \qquad \qquad \cdot 85321,6$$

$$-4n^2 + 3\frac{\mu}{a^3} \qquad \qquad -1\cdot 15842,8$$

$$\overline{\qquad -\cdot 30521,2}$$

$$a_{21} = \cdot 17605,6.$$

The equation for b_{21} is

$$b_{21} = \frac{2n}{n - 2n' + p} a_{21} + 4\frac{n - n'}{(n - 2n' + p)^2}(n - p)(1 + b_0) a_2$$

$$-2\frac{n - n'}{(n - 2n' + p)^2}(n - p) b_2 - 3\frac{n'^2}{(n - 2n' + p)^2}(1 + b_0),$$

$$\frac{2n}{n - 2n' + p} a_{21} \qquad \qquad \cdot 41201,8$$

$$4\,\frac{n - n'}{(n - 2n' + p)^2}(n - p)(1 + b_0) a_2 \qquad \cdot 03637,95$$

$$- 2\,\frac{n - n'}{(n - 2n' + p)^2}(n - p) b_2 \qquad - \cdot 02576,42$$

$$- 3\,\frac{n'^2}{(n - 2n' + p)^2}(1 + b_0) \qquad - \cdot 02308,03$$

$$b_{21} = \overline{\quad \cdot 39955,3 \quad}$$

The equation for a_{22} is

$$\left[(3n - 2n' + p)^2 - 4n^2 + 3\frac{\mu}{a^3}\right] a_{22} - 2(n - n')(n - p) a_2 + \frac{9}{2}\frac{\mu}{a^3} a_2$$

$$- 8\,\frac{n}{3n - 2n' - p}(n - n')(n - p)(1 + b_0) a_2 - 4(n - n')(n - p)(1 + b_0) b_2$$

$$- 4\,\frac{n}{3n - 2n' - p}(n - n')(n - p) b_2 - 3n'^2(1 + b_0)$$

$$- 6\,\frac{nn'^2}{3n - 2n' - p}(1 + b_0) = 0.$$

Here $\qquad - 2(n - n')(n - p) a_2 \qquad \qquad - \cdot 01545,45$

$$\frac{9}{2}\frac{\mu}{a^3} a_2 \qquad \qquad \cdot 03784,83$$

$$- 8\,\frac{n}{3n - 2n' - p}(n - n')(n - p)(1 + b_0) a_2 \qquad - \cdot 02181,13$$

$$- 4(n - n')(n - p)(1 + b_0) b_2 \qquad - \cdot 04415,05$$

$$- 4\,\frac{n}{3n - 2n' - p}(n - n')(n - p) b_2 \qquad - \cdot 01544,69$$

$$- 3n'^2(1 + b_0) \qquad \qquad - \cdot 01969,25$$

$$- 6\,\frac{nn'^2}{3n - 2n' - p}(1 + b_0) \qquad - \cdot 01383,78$$

$$\overline{\qquad \qquad - \cdot 09254,52}$$

$$(3n - 2n' - p)^2 \qquad \qquad 9 \cdot 46363,4$$

$$- 4n^2 + 3\frac{\mu}{a^3} \qquad \qquad - 1 \cdot 15842,8$$

$$\overline{\qquad \qquad 8 \cdot 30520,6}$$

$$a_{22} = \cdot 01114,30.$$

And

$$b_{22} = 2 \frac{n}{3n - 2n' - p} a_{22} + 4 \frac{n - n'}{(3n - 2n' - p)^2} (n - p)(1 + b_0) a_2$$

$$+ 2 \frac{n - n'}{(3n - 2n' - p)^2} (n - p) b_2 + 3 \frac{n'^2}{(3n - 2n' - p)^2} (1 + b_0),$$

$2 \dfrac{n}{3n - 2n' - p} a_{22}$	·00783,011
$4 \dfrac{n - n'}{(3n - 2n' - p)^2} (n - p)(1 + b_0) a_2$	·00327,988
$2 \dfrac{n - n'}{(3n - 2n' - p)^2} (n - p) b_2$	·00232,283
$3 \dfrac{n'^2}{(3n - 2n' - p)^2} (1 + b_0)$	·00208,086
$b_{22} =$	·01551,37

Second Approximation.   The complete equation for $n - p$ is

$$(n - p)^2 - 4n^2 + 3 \frac{\mu}{a^3} + 12 \frac{nn'^2}{n - p} b_2 (1 + b_0)$$

$$+ 2(n - n')(n - 2n' + p)[a_2 a_{21} - b_2 b_{21}] + \frac{9}{2} \frac{\mu}{a^3}[a_2 a_{21} + a_2 a_{22}]$$

$$+ \frac{3}{2} n'^2 [b_{21} + b_{22}] + 2(n - n')(3n - 2n' - p)[a_2 a_{22} - b_2 b_{22}]$$

$$+ 4 \frac{n}{n - p} (n - n')(n - 2n' - p)[b_2 a_{21} - a_2 b_{21}]$$

$$- 4 \frac{n}{n - p} (n - n')(3n - 2n' - p)[b_2 a_{22} - a_2 b_{22}]$$

$$+ 3 \frac{nn'^2}{n - p}[b_{21} - b_{22}] = 0.$$

$-4n^2 + 3 \dfrac{\mu}{a^3}$	$-1·15842,8$
$12 \dfrac{nn'^2}{n - p} b_2 (1 + b_0)$	·00080,8
$2(n - n')(n - 2n' + p)[a_2 a_{21} - b_2 b_{21}]$	$- ·00520,1$

$$\frac{9}{2}\frac{\mu}{a^3}[a_2 a_{21} + a_2 a_{22}] \qquad\qquad \cdot 00708,4$$

$$\frac{3}{2} n'^2 [b_{21} + b_{22}] \qquad\qquad \cdot 00406,9$$

$$2(n-n')(3n-2n'-p)[a_2 a_{22} - b_2 b_{22}] \qquad - \cdot 00048,2$$

$$4\frac{n}{n-p}(n-n')(n-2n'+p)[b_2 a_{21} - a_2 b_{21}] \qquad - \cdot 00397,5$$

$$-4\frac{n}{n-p}(n-n')(3n-2n'-p)[b_2 a_{22} - a_2 b_{22}] \qquad - \cdot 00003,0$$

$$3\frac{nn'^2}{n-p}[b_{21} - b_{22}] \qquad\qquad \cdot 00756,1$$

$$(n-p)^2 = \quad \overline{1 \cdot 14859,4}$$

$$n-p = 1 \cdot 07172,5$$
$$p = \quad \cdot 00912,4$$
$$p : n = \quad \cdot 00844,2.$$

Apply these numbers in the equation for $1 + b_0$:—

$$1+b_0 = \frac{n}{n-p} - 3\frac{n'^2}{(n-p)^2}b_2(1+b_0) - \frac{n-n'}{(n-p)^2}(n-2n'+p)[b_2 a_{21} - a_2 b_{21}]$$

$$+ \frac{n-n'}{(n-p)^2}(3n-2n'-p)[b_2 a_{22} - a_2 b_{22}] - \frac{3}{4}\frac{n'^2}{(n-p)^2}[b_{21} - b_{22}],$$

$$\frac{n}{n-p} \qquad\qquad 1 \cdot 00851,33$$

$$-3\frac{n'^2}{(n-p)^2}b_2(1+b_0) \qquad - \cdot 00017,51$$

$$-\frac{n-n'}{(n-p)^2}(n-2n'+p)[b_2 a_{21} - a_2 b_{21}] \qquad \cdot 00086,55$$

$$\frac{n-n'}{(n-p)^2}(3n-2n'-p)[b_2 a_{22} - a_2 b_{22}] \qquad \cdot 00000,64$$

$$-\frac{3}{4}\frac{n'^2}{(n-p)^2}[b_{21} - b_{22}] \qquad - \cdot 00163,92$$

$$1 + b_0 = \quad \overline{1 \cdot 00757,1}$$

Continuing the approximation for a_{21}, b_{21}, a_{22}, b_{22} the various terms found are the following:—

$$
\begin{array}{ll}
·01538,88 & \text{divisor} \\
·03784,83 & ·86169,5 \\
-·07221,53 & -1·15842,8 \\
-·04410,93 & \overline{-·29673,3} \\
·05097,33 & \\
·01975,81 & \\
·04601,14 & a_{21} = ·18082,0 \\
\overline{·05365,53} &
\end{array}
\qquad
\begin{array}{ll}
-·01538,88 & \text{divisor} \\
·03784,83 & 9·43549,4 \\
-·02182,34 & -1·15842,8 \\
-·04410,93 & \overline{8·27706,6} \\
-·01540,41 & \\
-·01975,82 & \\
-·01390,46 & a_{22} = ·01118,03 \\
\overline{-·09254,01} &
\end{array}
$$

$$
\begin{array}{rl}
& ·42108,0 \\
& ·03598,79 \\
- & ·02540,22 \\
- & ·02292,94 \\
b_{21} = & \overline{·40873,6}
\end{array}
\qquad
\begin{array}{rl}
& ·00786,805 \\
& ·00328,659 \\
& ·00231,985 \\
& ·00209,403 \\
b_{22} = & \overline{·01556,85}
\end{array}
$$

Third Approximation.   We find

$$
\begin{array}{rl}
& 1·15842,8 \\
- & ·00081,40 \\
& ·00535,02 \\
- & ·00723,54 \\
- & ·00416,02 \\
& ·00048,35 \\
& ·00407,40 \\
& ·00002,97 \\
- & ·00777,57 \\
(n-p)^2 = & \overline{1·14838,0}
\end{array}
\qquad
\begin{array}{rl}
n - p = & 1·07162,5 \\
p = & ·00922,4 \\
p : n = & ·00853,5 \\
\end{array}
\qquad
\begin{array}{rl}
& 1·00860,75 \\
- & ·00017,57 \\
& ·00087,94 \\
& ·00000,64 \\
- & ·00167,84 \\
1 + b_0 = & \overline{1·00763,9}
\end{array}
$$

$$
\begin{array}{ll}
·01538,73 & \text{divisor} \\
·03784,83 & ·86188,04 \\
-·07220,54 & -1·15842,8 \\
-·04410,81 & \overline{-·29654,8} \\
·05096,30 & \\
·01975,95 & \\
·04600,96 & \\
\overline{·05365,42} & a_{21} = ·18092,9.
\end{array}
\qquad
\begin{array}{ll}
-·01538,73 & \text{divisor} \\
·03784,83 & 9·43488,0 \\
-·02182,35 & -1·15842,8 \\
-·04410,81 & \overline{8·27645,2} \\
-·01540,32 & \\
-·01975,95 & \\
-·01390,60 & \\
\overline{-·09253,93} & a_{22} = ·01118,10.
\end{array}
$$

$$
\begin{array}{rl}
& ·42128,90 \\
& ·03597,92 \\
- & ·02539,43 \\
- & ·02292,60 \\
b_{21} = & \overline{·40894,8}
\end{array}
\qquad
\begin{array}{rl}
& ·00786,881 \\
& ·00328,671 \\
& ·00231,978 \\
& ·00209,430 \\
b_{22} = & \overline{·01556,96}
\end{array}
$$

Fourth Approximation.

$1\cdot15842,8$	$1\cdot00862,79$
$-\ \cdot00081,41$	$-\ \cdot00017,57$
$\cdot00534,22 \quad n-p=1\cdot07160,3$	$\cdot00088,04$
$-\ \cdot00727,13 \qquad p=\ \cdot00924,6$	$\cdot00000,64$
$-\ \cdot00416,24 \quad p:n=\ \cdot00855,4$	$-\ \cdot00167,94$
$\cdot00048,35$	$1+b_0=\ \overline{1\cdot00766,0}$
$\cdot00407,66$	
$\cdot00002,97$	
$-\ \cdot00778,05$	
$(n-p)^2=\ \overline{1\cdot14833,2}$	

The values already found for the remaining quantities are sufficiently exact.

These numbers give, taking after Hansen,

$$e\,(1+b_0)=\cdot05491$$

$$e=\cdot05449$$

$$2\,(1+b_0)\,e=\cdot10982,0=22651'''\!\cdot9$$

$$a_{21}e=\cdot00986,03$$

$$b_{21}e=\cdot02228,44=4596'''\!\cdot6$$

$$a_{22}e=\cdot00060,93$$

$$b_{22}e=\cdot00084,85=\ 175'''\!\cdot1,$$

and taking the Moon's mean annual motion $17325593''$, the annual motion of the apse is

$$148202''=41°\,10'\,2''.$$

LECTURE XIII.

THE MOTION OF THE APSE, AND THE CHANGE OF THE ECCENTRICITY.

WE have seen that when the eccentricity of the Moon's orbit is not considered we may write

$$\frac{1}{r} = \frac{1}{a}\left[1 + a_2 \cos 2\left(\theta - \theta'\right)\right],$$

$$H = na^2\left[1 + h_2 \cos 2\left(\theta - \theta'\right)\right],$$

where $\quad a_2 = \dfrac{3}{2}\, m^2 \cdot \dfrac{2-m}{1-m} \cdot \dfrac{1}{3 - 8m + \dfrac{11}{2}\, m^2}; \quad h_2 = \dfrac{3}{4}\, \dfrac{m^2}{1-m}.$

Let us introduce the two new arbitraries e, ϖ by writing

$$H = hna^2\left[1 + h_2 \cos\left(2 - 2m\right)\theta\right],$$

$$\frac{1}{r} = \frac{1}{h^2 a}\left[1 + a_2 \cos\left(2 - 2m\right)\theta + e \cos\left(\theta - \varpi\right)\right],$$

where h is a third arbitrary, which may be chosen to suit our convenience; it must be unity when $e = 0$.

Then

$$\frac{dH}{d\theta} = \frac{dH}{dt} \cdot \frac{dt}{d\theta} = -\frac{3}{2}\, m^2 n^2 r^2 \sin 2\left(\theta - \theta'\right) \frac{dt}{d\theta}$$

$$= -\frac{3}{2}\, m^2 n^2 \frac{r^4}{H}\left[\sin\left(2 - 2m\right)\theta + 4me \cos\left(2 - 2m\right)\theta \sin\left(\theta - \varpi\right)\right]$$

$$= -\frac{3}{2}\, m^2 n^2 \frac{h^8 a^4}{hna^2}\left[1 - \left(4a_2 + h_2\right)\cos\left(2 - 2m\right)\theta - 4e \cos\left(\theta - \varpi\right)\right]$$

$$\times \left[\sin\left(2 - 2m\right)\theta + 4me \cos\left(2 - 2m\right)\theta \sin\left(\theta - \varpi\right)\right],$$

and also

$$\frac{dH}{d\theta} = \frac{dh}{d\theta} na^2 [1 + h_2 \cos(2 - 2m)\theta] - na^2 h(2 - 2m) h_2 \sin(2 - 2m)\theta.$$

Now we may put $h = 1 + \eta$, where η vanishes with e. Neglecting powers of e above the first

$$\frac{d\eta}{d\theta} = \frac{dh}{d\theta} = 3m^2(1 + m)e\sin(\overline{1 - 2m}\theta + \varpi)$$
$$+ 3m^2(1 - m)e\sin(\overline{3 - 2m}\theta - \varpi) - 9m^2\eta\sin(2 - 2m)\theta.$$

Neglect at first the last term:

$$\eta = -3m^2\frac{1 + m}{1 - 2m}e\cos(\overline{1 - 2m}\theta + \varpi) - 3m^2\frac{1 - m}{3 - 2m}e\cos(\overline{3 - 2m}\theta - \varpi).$$

Substitute this in the last term, and we get

$$\frac{d\eta}{d\theta} = 3m^2(1 + m)e\sin(\overline{1 - 2m}\theta + \varpi) + 3m^2(1 - m)e\sin(\overline{3 - 2m}\theta - \varpi)$$
$$+ \frac{27}{2}m^4\frac{2 + 4m - 4m^2}{(1 - 2m)(3 - 2m)}e\sin(\theta - \varpi).$$

Now consider the other equation

$$\frac{d^2r}{dt^2} = \frac{H^2}{r^3} - \frac{\mu}{r^2} + \frac{1}{2}m^2n^2r + \frac{3}{2}m^2n^2r\cos 2(\theta - \theta')$$
$$= \frac{H^2}{r^3} - \frac{\mu}{r^2} + \frac{1}{2}m^2n^2r + \frac{3}{2}m^2n^2r[\cos(2 - 2m)\theta$$
$$- 2me\cos(\overline{1 - 2m}\theta + \varpi) + 2me\cos(\overline{3 - 2m}\theta - \varpi)].$$

Differentiate the assumed expression for $\frac{1}{r}$, and let h be chosen so that the first differential coefficient shall have the same form as if h, e, ϖ were constant.

Thus

$$\frac{1}{r^2}\frac{dr}{dt} = \frac{1}{h^2a}(2 - 2m)a_2\sin(2 - 2m)\theta\frac{d\theta}{dt} + \frac{1}{h^2a}e\sin(\theta - \varpi)\frac{d\theta}{dt},$$

where

$$-\frac{2}{h}\frac{dh}{d\theta}[1 + a_2\cos(2 - 2m)\theta] + \frac{de}{d\theta}\cos(\theta - \varpi) + e\frac{d\varpi}{d\theta}\sin(\theta - \varpi) = 0,$$

or

$$\frac{de}{d\theta}\cos(\theta-\varpi)+e\frac{d\varpi}{d\theta}\sin(\theta-\varpi)=6m^2(1+m)\,e\sin(\overline{1-2m}\,\theta+\varpi)$$

$$+\quad 6m^2(1-m)\,e\sin(\overline{3-2m}\,\theta-\varpi)$$

$$+\left\{27m^4\frac{2+4m-4m^2}{(1-2m)(3-2m)}-6m^3a_2\right\}e\sin(\theta-\varpi),$$

and

$$\frac{dr}{dt}=\frac{H}{h^2a}(2-2m)\,a_2\sin(2-2m)\theta+\frac{H}{h^2a}e\sin(\theta-\varpi)$$

$$=\frac{na}{h}(2-2m)\,a_2\sin(2-2m)\,\theta$$

$$+\frac{na}{h}\left[1+h_2\cos(2-2m)\theta\right]e\sin(\theta-\varpi),$$

$$\frac{d^2r}{dt^2}=\frac{na}{h}(2-2m)^2\,a_2\cos(2-2m)\,\theta\frac{d\theta}{dt}$$

$$-\frac{na}{h}(2-2m)\,h_2\sin(2-2m)\,\theta\,e\sin(\theta-\varpi)\frac{d\theta}{dt}$$

$$+\frac{na}{h}\left[1+h_2\cos(2-2m)\,\theta\right]e\cos(\theta-\varpi)\frac{d\theta}{dt}$$

$$-\frac{na}{h^2}\frac{dh}{d\theta}(2-2m)\,a_2\sin(2-2m)\,\theta\frac{d\theta}{dt}$$

$$+\frac{na}{h}\left[1+h_2\cos(2-2m)\theta\right]\left[\frac{de}{d\theta}\sin(\theta-\varpi)-e\frac{d\varpi}{d\theta}\cos(\theta-\varpi)\right]\frac{d\theta}{dt}.$$

Multiply by r^2/n^2a^3; then since

$$r^2\frac{d\theta}{dt}=H=hna^2\left[1+h_2\cos(2-2m)\,\theta\right],$$

we have

$$\frac{r^2}{n^2a^3}\frac{d^2r}{dt^2}=(2-2m)^2\,a_2\cos(2-2m)\,\theta$$

$$-(2-2m)\,h_2\sin(2-2m)\,\theta\,e\sin(\theta-\varpi)$$

$$+e\cos(\theta-\varpi)+2h_2\cos(2-2m)\,\theta\,e\cos(\theta-\varpi)$$

$$-\frac{1}{h}\frac{dh}{d\theta}(2-2m)\,a_2\sin(2-2m)\,\theta$$

$$+[1+h_2\cos(2-2m)\theta]^2\left[\frac{de}{d\theta}\sin(\theta-\varpi)-e\frac{d\varpi}{d\theta}\cos(\theta-\varpi)\right],$$

and this is equal to

$$\frac{H^2}{n^2a^3r} - \frac{\mu}{n^2a^3} + \frac{1}{2}m^2\frac{r^3}{a^3} + \frac{3}{2}m^2\frac{r^3}{a^3}\left[\cos(2-2m)\theta\right.$$

$$\left. - 2me\cos(\overline{1-2m}\theta+\varpi) + 2me\cos(\overline{3-2m}\theta-\varpi)\right]$$

$$= \left[1 + \frac{h_2^2}{2} + 2h_2\cos(2-2m)\theta\right]\left[1 + a_2\cos(2-2m)\theta + e\cos(\theta+\varpi)\right]$$

$$- \frac{\mu}{n^2a^3} + \frac{1}{2}m^2h^6\left[1 - 3a_2\cos(2-2m)\theta - 3e\cos(\theta-\varpi)\right]$$

$$+ \frac{3}{2}m^2h^6\left[\cos(2-2m)\theta + 6a_2e\cos(\theta-\varpi)\right.$$

$$\left. - \left(\frac{3}{2}+2m\right)e\cos(\overline{1-2m}\theta+\varpi) - \left(\frac{3}{2}-2m\right)e\cos(\overline{3-2m}\theta-\varpi)\right].$$

The terms in these two expressions which are independent of e give no new information; equating the others:—

$$-(2-2m)h_2e\sin(2-2m)\theta\sin(\theta-\varpi)$$

$$- \frac{1}{h}\frac{dh}{d\theta}(2-2m)a_2\sin(2-2m)\theta$$

$$+ \left[1 + h_2\cos(2-2m)\theta\right]^2\left[\frac{de}{d\theta}\sin(\theta-\varpi) - e\frac{d\varpi}{d\theta}\cos(\theta-\varpi)\right]$$

$$= 3m^2\eta - \frac{3}{2}m^2e\cos(\theta-\varpi) + \frac{3}{2}m^2\left[6a_2e\cos(\theta-\varpi)\right.$$

$$\left. - \left(\frac{3}{2}+2m\right)e\cos(\overline{1-2m}\theta+\varpi) - \left(\frac{3}{2}-2m\right)e\cos(\overline{3-2m}\theta-\varpi)\right]$$

$$+ 9m^2\eta\cos(2-2m)\theta.$$

Reducing this expression

$$-\frac{de}{d\theta}\sin(\theta-\varpi) + e\frac{d\varpi}{d\theta}\cos(\theta-\varpi) = \left(\frac{3}{2}m^2 - \frac{3}{8}m^4\right)e\cos(\theta-\varpi)$$

$$+ \left(\frac{3}{2}m^2 + 3m^3 + \frac{63}{8}m^4\right)e\cos(\overline{1-2m}\theta+\varpi)$$

$$+ \left(3m^2 - 3m^3 + \frac{15}{8}m^4\right)e\cos(\overline{3-2m}\theta-\varpi)$$

and from before

$$\frac{de}{d\bar\theta}\cos(\theta-\varpi)+e\frac{d\varpi}{d\theta}\sin(\theta-\varpi)=18m^4e\sin(\theta-\varpi)$$
$$+(6m^2+6m^3)\,e\sin(\overline{1-2m}\theta+\varpi)$$
$$+(6m^2-6m^3)\,e\sin(\overline{3-2m}\theta+\varpi).$$

Hence,

$$\frac{de}{d\theta}=\left(-\frac{3}{4}\,m^2\qquad+\frac{147}{16}\,m^4\right)e\sin 2\,(\theta-\varpi)$$

$$+\left(\quad\frac{27}{4}\,m^2-3m^3-\quad3\,\,m^4\right)e\sin(2-2m)\,\theta$$

$$+\left(-\frac{15}{4}\,m^2-\frac{9}{2}\,m^3-\frac{63}{16}\,m^4\right)e\sin(2m\theta-2\varpi)$$

$$+\left(\quad\frac{3}{2}\,m^2-\frac{3}{2}\,m^3-\frac{15}{16}\,m^4\right)e\sin(\overline{4-2m}\theta-2\varpi),$$

$$\frac{d\varpi}{d\theta}=\frac{3}{4}\,m^2+\frac{141}{16}\,m^4+\left(\quad\frac{3}{4}\,m^2\qquad-\frac{147}{16}m^4\right)\cos 2\,(\theta-\varpi)$$

$$+\left(\quad\frac{9}{4}\,m^2-6\,m^3+\frac{39}{8}\,m^4\right)\cos(2-2m)\,\theta$$

$$+\left(\quad\frac{15}{4}m^2+\frac{9}{2}\,m^3+\frac{63}{16}\,m^4\right)\cos(2m\theta-2\varpi)$$

$$+\left(-\frac{3}{2}\,m^2+\frac{3}{2}\,m^3+\frac{15}{16}\,m^4\right)\cos(\overline{4-2m}\theta-2\varpi).$$

We notice that among these terms, one is of long period, approximately semiannual, and will become of greater relative importance than the others on integration.

To effect this integration, assume

$$\Pi=\varpi+\alpha\sin 2\,(\theta-\varpi)+\beta\sin(2-2m)\,\theta+\gamma\sin\{(4-2m)\theta-2\varpi\},$$

so that the mean motion of Π is the same as that of ϖ, and substitute in the equation.

Then

$$\frac{d\Pi}{d\theta} = \quad \frac{3}{4}\,m^2 \qquad + \frac{141}{16}\,m^4 \quad - \frac{3}{4}\,m^2\alpha \qquad + \frac{3}{2}\,m^2\gamma$$

$$+ \left[\ \ \frac{3}{4}\,m^2 \qquad - \frac{147}{16}\,m^4 + 2\alpha - \frac{3}{2}\,m^2\alpha + \frac{15}{4}\,m^4\beta - \frac{9}{4}\,m^2\gamma\ \right]$$
$$\times \cos 2\,(\theta - \varpi)$$

$$+ \left[\ \ \frac{9}{4}\,m^2 - 6m^3 + \frac{39}{8}\,m^4 + (2 - 2m)\,\beta - 6m^2\alpha - \frac{3}{4}\,m^3\gamma\ \right]$$
$$\times \cos (2 - 2m)\,\theta$$

$$+ \left[\ \ \frac{15}{4}\,m^2 + \frac{9}{2}\,m^3 + \frac{63}{16}\,m^4 - \frac{9}{4}\,m^2\alpha\ \right] \cos (2m\theta - 2\Pi)$$

$$+ \left[\ -\frac{3}{2}\,m^2 + \frac{3}{2}\,m^3 + \frac{15}{16}\,m^4 + (4 - 2m)\,\gamma - \frac{9}{4}\,m^2\iota - \frac{3}{2}\,m^2\gamma\ \right]$$
$$\times \cos \{(4 - 2m)\,\theta - 2\varpi\},$$

so that if we take

$$\alpha = -\frac{3}{8}\,m^2 \qquad\qquad + \frac{219}{32}\,m^4,$$

$$\beta = -\frac{9}{8}\,m^2 + \frac{15}{8}\,m^3 - \frac{99}{64}\,m^4,$$

$$\gamma = \quad\frac{3}{8}\,m^2 - \frac{3}{16}\,m^3 - \frac{51}{128}\,m^4,$$

we have

$$\frac{d\Pi}{d\theta} = \frac{3}{4}\,m^2 + \frac{309}{32}\,m^4 + \left[\frac{15}{4}\,m^2 + \frac{9}{2}\,m^3 + \frac{45}{8}\,m^4\right] \cos (2m\theta - 2\Pi).$$

If we write $\quad\quad\quad m\theta - \Pi = \psi$,

this becomes $\quad\quad\quad \dfrac{d\psi}{d\theta} = a - b\cos 2\psi$,

where

$$a = m - \frac{3}{4}\,m^2 - \frac{309}{32}\,m^4, \quad b = \frac{15}{4}\,m^2 + \frac{9}{2}\,m^3 + \frac{45}{8}\,m^4,$$

and the solution is

$$\tan^{-1}\left(\sqrt{\frac{a+b}{a-b}}\,\tan\psi\right) = \theta\sqrt{a^2 - b^2} + \text{constant}.$$

Hence if we denote by $\dfrac{d\varpi_0}{d\theta}$ the mean rate of change of ϖ, we have

$$m - \frac{d\varpi_0}{d\theta} = \sqrt{a^2 - b^2}$$

$$= m - \frac{3}{4} m^2 - \frac{225}{32} m^3 - \frac{4071}{128} m^4,$$

or $$\frac{d\varpi_0}{d\theta} = \frac{3}{4} m^2 + \frac{225}{32} m^3 + \frac{4071}{128} m^4.$$

We observe that $a + b$ and $a - b$ are the rates of separation of the Sun from the apse when the Sun and the apse are at quadratures and syzygies with one another, respectively,—that is if we take Π for the longitude of the apse, or, what is the same thing, if we ignore small terms of short period. Hence the mean rate of separation of the Sun from the apse is a mean proportional between its rates when at quadratures and syzygies respectively with the apse*.

[* This is the analogue for the case of the apse of Machin and Pemberton's theorem on the motion of the node, inserted in the third edition of the *Principia* as a scholium to prop. xxxiii., lib. iii. See some notes by Adams in Brewster's *Life of Newton*, Appendix xxx.]

LECTURE XIV.

THE LATITUDE AND THE MOTION OF THE NODE.

LET us first treat this problem on the supposition that the latitude is so small that its square may be neglected. The equation of motion, taken from Lecture II, may be written

$$\frac{d^2z}{dt^2} = -\frac{z}{r}\left[\frac{\mu}{r^2} + \frac{m'r}{r'^3}\left(1 + \frac{E-M}{E+M}\frac{r}{r'}3\cos\omega\right)\right],$$

where $z = r\sin(\text{latitude})$ and the cube of s is omitted; or neglecting the parallactic terms

$$\frac{d^2z}{dt^2} = -z\left[\frac{\mu}{r^3} + \frac{m'}{r'^3}\right].$$

The value of μ/r^3 may be considered known by the operations which have determined the motion in an orbit coinciding with the ecliptic; that is to say,

$$\frac{\mu}{r^3} = \frac{\mu}{a^3}\left[1 + \frac{3}{2}a_2{}^2 + 3a_2\cos 2\psi + \left(\frac{3}{2}a_2{}^2 + 3a_4\right)\cos 4\psi\right],$$

where a has the definition of Lectures IV, V; or numerically, taking

$$n - n' = 1,$$

$$\frac{\mu}{r^3} = 1\cdot17150,3 + \cdot02523,0\cos 2t + \cdot00025,15\cos 4t.$$

And

$$\frac{m'}{r'^3} = n'^2 = \cdot00653,6.$$

Hence

$$\frac{d^2z}{dt^2} = -z\left[1\cdot17803,9 + \cdot02523,0\cos 2t + \cdot00025,15\cos 4t\right].$$

Let us now consider the equation

$$\frac{d^2z}{dt^2} + Pz = 0,$$

where $\qquad P = q_0 + 2q_1 \cos 2t + 2q_2 \cos 4t,$

in which q_1, q_2 are supposed small.

Suppose a term in z to be $c \cos(kt + \beta)$; when this is substituted in Pz there will arise terms

$$c \cos(\overline{k - 2}t + \beta) \qquad\qquad c \cos(\overline{k + 2}t + \beta)$$
$$c \cos(\overline{k - 4}t + \beta) \qquad\qquad c \cos(\overline{k + 4}t + \beta).$$

Let us therefore assume

$$z = c\,[\cos(kt + \beta) + c_1 \cos(\overline{k + 2}t + \beta) + c_2 \cos(\overline{k + 4}t + \beta) + \ldots$$
$$+ c_{-1} \cos(\overline{k - 2}t + \beta) + c_{-2} \cos(\overline{k - 4}t + \beta) + \ldots]$$

c is arbitrary; we have to determine k, c_1, c_{-1}, &c.

Substitute and equate coefficients:

$$\ldots + [-(k-4)^2 + q_0]c_{-2} + q_1 c_{-1} + q_2 + \ldots \qquad\qquad = 0$$
$$\ldots\ldots + q_1 c_{-2} + [-(k-2)^2 + q_0]c_{-1} + q_1 + q_2 c_1 + \ldots \qquad = 0$$
$$\ldots\ldots + q_2 c_{-2} + q_1 c_{-1} + [-k^2 + q_0] + q_1 c_1 + q_2 c_2 + \ldots \qquad = 0$$
$$\ldots\ldots + q_2 c_{-1} + q_1 + [-(k+2)^2 + q_0]c_1 + q_1 c_2 + \ldots = 0$$
$$\ldots\ldots + q_2 + q_1 c_1 + [-(k+4)^2 + q_0]c_2 + \ldots = 0.$$

If q_1, q_2,... are neglected, we have simply

$$-k^2 + q_0 = 0;$$

this is a first approximation to the value of k.

Taking q_1 into account and neglecting q_2

$$c_{-1} = -\frac{q_1}{q_0 - (k-2)^2},$$

$$c_1 = -\frac{q_1}{q_0 - (k+2)^2}.$$

In the actual case considered we notice that q_0 does not differ widely from unity. Hence k is nearly equal to unity also, and the denominator in c_{-1} is small, and makes c_{-1} much more important than c_1.

If we substitute these values in the third equation above, we have

$$k^2 - q_0 + q_1{}^2 \left\{ \frac{1}{q_0 - (k+2)^2} + \frac{1}{q_0 - (k-2)^2} \right\} = 0,$$

whence

$$(k^2 - q_0)^3 - 8(k^2 - q_0)^2 - \{16(q_0 - 1) + 2q_1{}^2\}(k^2 - q_0) - 8q_1{}^2 = 0,$$

which may be put under the form

$$(k^2 - q_0)^2 + 2(q_0 - 1)(k^2 - q_0) = -q_1{}^2 + \frac{1}{4}q_1{}^2(k^2 - q_0) + \frac{1}{8}(k^2 - q_0)^3,$$

whence

$$(k^2 - 1)^2 = (q_0 - 1)^2 - q_1{}^2 - \frac{1}{4}q_1{}^2(k^2 - q_0) + \frac{1}{8}(k^2 - q_0)^3.$$

With this equation we can approximate very rapidly to the value of k. Taking as a first approximation

$$k^2 - q_0 = 0,$$

substitute this value of k in the small terms and we get as a second approximation

$$k = 1 \cdot 08516,9.$$

Whence the ratio of the retrograde motion of the node to the Moon's mean motion is

$$k/n - 1 = g - 1 = \cdot 00399,7,$$

where g is written for k/n. This value is very correct. Taking the Moon's mean annual motion as $17325593''$, the resulting annual retrograde motion of the node is

$$69252'' = 19° \, 14' \, 12''.$$

Next find the values of the coefficients c_{-1}, c_1, c_{-2}, c_2. We have

$$q_1 = \quad \cdot 01261,5,$$
$$q_0 - (k - 2)^2 = \quad \cdot 34112,3,$$
$$q_0 - (k + 2)^2 = - \, 8 \cdot 34022,8 \, ;$$

whence as a first approximation

$$c_{-1} = - \, \cdot 03698,19, \qquad c_1 = \cdot 00151,26.$$

Hence

$$q_1 c_{-1} + q_2 = -\cdot00034{,}02, \qquad q_1 c_1 + q_2 = \cdot00014{,}49,$$

and

$$q_0 - (k-4)^2 = -7\cdot31821, \qquad q_0 - (k+4)^2 = -24\cdot6809,$$

whence

$$c_{-2} = -\cdot00004{,}650,$$
$$c_2 = \cdot00000{,}587.$$

A second approximation to c_{-1}, c_1 gives

$$[-(k-2)^2 + q_0]\, c_{-1} = -(q_1 + q_2 c_1 + q_1 c_{-2}),$$
$$[-(k+2)^2 + q_0]\, c_1 = -(q_1 + q_2 c_{-1} + q_1 c_2);$$

with the above values

$$q_1 + q_2 c_1 + q_1 c_{-2} = \cdot01261{,}47, \qquad q_1 + q_2 c_{-1} + q_1 c_2 = \cdot01261{,}03,$$

so that

$$c_{-1} = -\cdot03698{,}00,$$
$$c_1 = \cdot00151{,}20.$$

LECTURE XV.

LET us consider the change in the plane of the orbit produced in an indefinitely small time dt by the action of a given disturbing force. Let Z be the resolved part of the disturbing force at any time in a direction perpendicular to the plane in which the body is moving at the instant. Imagine the force Z to act by impulses at the small intervals of time dt, then $Z dt$ will be the indefinitely small velocity generated by the force Z in the time dt, in the direction perpendicular to the plane of orbit at the instant.

Let FP be the radius vector and P the position of the body at the instant. Also let PT represent the velocity at

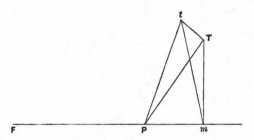

the instant in magnitude and direction; then if Tt be taken perpendicular to the plane FPT and equal to $Z dt$, the velocity and its direction after the impulse will be represented by Pt, and the new plane of the orbit by FPt. Draw Tm perpendicular to FP and join tm; then tmT is the angle through which the plane of the orbit has been turned about the radius vector FP in the indefinitely short time dt.

Now
$$tmT = \frac{tT}{Tm} = \frac{Zdt}{v},$$

where v is the resolved part of the velocity at P perpendicular to the radius vector. But
$$H = vr;$$

hence the angle through which the orbit is turned in an indefinitely short time dt is
$$\frac{rZ}{H} dt.$$

To find the corresponding changes in the elements that determine the plane of the orbit, namely, the inclination of the orbit to a fixed plane, and the longitude of the node on that plane. Let NPQ be the great circle which represents the plane of the

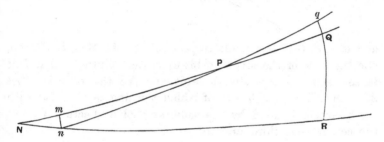

orbit at the time t, NR the plane of reference, usually the plane of the ecliptic, P the position of the body at the same time, and let $i = PNR$, the inclination, and let N be the longitude of the node.

Let nPq be the position of the orbit at time $t + dt$.

Take $NQ = 90°$; draw nm perpendicular to NPQ and qQR perpendicular to NR; then $QR = i$; and by what we have just proved
$$NPn = \frac{Zr}{H} dt.$$

Therefore $nm = \dfrac{Zr}{H} dt . \sin \theta,$ $qQ = \dfrac{Zr}{H} dt . \cos \theta,$

where $\theta = nP.$

But $nm = Nn \sin i = \sin i \, dN;$ $qQ = di.$

Therefore

$$\frac{di}{dt} = \frac{Zr \cos \theta}{H},$$

$$\frac{dN}{dt} = \frac{Zr \sin \theta}{H \sin i},$$

which give the changes of the elements required.

Now let NMS be a spherical triangle, the centre of the sphere being G, the centre of gravity of the Earth and Moon;

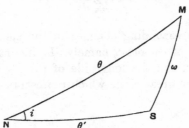

and let GS, GM, GN point respectively to the Sun, the Moon, and the node of the Moon's orbit upon the ecliptic, so that NM is the plane of the Moon's orbit and NS the ecliptic. Let $MS = \omega$, $NS = \theta'$, $NM = \theta$, of which the first is identical with the quantity denoted by the same symbol in Lecture II, but the second and third are not so.

Then, following Lecture II, the forces on the Moon are

$$\frac{\mu}{r^2} + \frac{m'r}{r'^3} \qquad \text{in } MG,$$

$$-\frac{m'r}{r'^3} 3 \cos \omega, \quad \text{in } SG,$$

if we ignore the parallactic terms.

This latter may be resolved into

$-\dfrac{m'r}{r'^3} 3 \cos \omega \times (\cos \theta \cos \theta' + \sin \theta \sin \theta' \cos i)$ in MG,

$\dfrac{m'r}{r'^3} 3 \cos \omega \times (\sin \theta \cos \theta' - \cos \theta \sin \theta' \cos i)$ perpendicular to MG in the plane of the orbit,

$\dfrac{m'r}{r'^3} 3 \cos \omega \times \sin \theta' \sin i$ perpendicular to the plane of the orbit.

Now

$$\cos \omega = \cos \theta \cos \theta' + \sin \theta \sin \theta' \cos i$$
$$= \cos (\theta - \theta') \cos^2 \frac{i}{2} + \cos (\theta + \theta') \sin^2 \frac{i}{2},$$
$$\sin \theta \cos \theta' - \cos \theta \sin \theta' \cos i$$
$$= \sin (\theta - \theta') \cos^2 \frac{i}{2} + \sin (\theta + \theta') \sin^2 \frac{i}{2}.$$

Hence we have the following expressions for the three forces:

$$P = \frac{\mu}{r^2} + \frac{m'r}{r'^3} - \frac{3}{2} \frac{m'r}{r'^3} \left[\{1 + \cos 2(\theta - \theta')\} \cos^4 \frac{i}{2} + \{\cos 2\theta + \cos 2\theta'\} \right.$$
$$\left. \times 2 \cos^2 \frac{i}{2} \sin^2 \frac{i}{2} + \{1 + \cos 2(\theta - \theta')\} \sin^4 \frac{i}{2} \right],$$

$$T = \frac{3}{2} \frac{m'r}{r'^3} \left[\sin 2(\theta - \theta') \cos^4 \frac{i}{2} \right.$$
$$\left. + \sin 2\theta . 2 \cos^2 \frac{i}{2} \sin^2 \frac{i}{2} + \sin 2(\theta + \theta') \sin^4 \frac{i}{2} \right],$$

$$Z = \frac{3}{2} \frac{m'r}{r'^3} \sin i \left[- \sin (\theta - 2\theta') \cos^2 \frac{i}{2} \right.$$
$$\left. + \sin \theta \cos i + \sin (\theta + 2\theta') \sin^2 \frac{i}{2} \right].$$

Now we have seen

$$\frac{di}{dt} = - \frac{Zr \cos \theta}{H}, \qquad \frac{dN}{dt} = - \frac{Zr \sin \theta}{H \sin i};$$

also, the rate of advance of the node along the orbit is

$$- \frac{Zr \sin \theta}{H \tan i}.$$

Thus the equations of motion become

$$\frac{H}{r^2} = \frac{d\theta}{dt} - \frac{Zr \sin \theta}{H \tan i},$$

together with
$$\frac{d^2r}{dt^2} - \frac{H^2}{r^3} = - P,$$

$$\frac{dH}{dt} = - rT.$$

LECTURE XVI.

MOTION IN AN ORBIT OF ANY INCLINATION (*continued*).

To satisfy the equation at the end of Lecture XV, assume

$$r = a\left[1 + A_1 \cos 2(\theta - \theta') + A_2 \cos 2\theta + A_3 \cos 2\theta' + A_4 \cos 2(\theta + \theta')\right],$$

neglecting the square of the disturbing force and the eccentricity; thus in the small terms we write

$$r = a, \quad \frac{d\theta}{dt} = n, \quad r' = a', \quad \frac{d\theta'}{dt} = n'.$$

Hence

$$-\frac{d^2 r}{dt^2} = n^2 a\left[(2-2m)^2 A_1 \cos 2(\theta - \theta') + 4A_2 \cos 2\theta \right.$$
$$\left. + 4m^2 A_3 \cos 2\theta' + (2+2m)^2 A_4 \cos 2(\theta + \theta')\right];$$

substitute in the equation

$$H^2 = r^3 P + r^3 \frac{d^2 r}{dt^2};$$

therefore

$$H^2 = \mu a\left[1 + A_1 \cos 2(\theta - \theta') + A_2 \cos 2\theta + A_3 \cos 2\theta' \right.$$
$$\left. + A_4 \cos 2(\theta + \theta')\right]$$
$$+ m^2 n^2 a^4$$
$$- \frac{3}{2} m^2 n^2 a^4 \left[\{1 + \cos 2(\theta - \theta')\} \cos^4 \frac{i}{2}\right.$$
$$+ \{\cos 2\theta + \cos 2\theta'\} 2 \cos^2 \frac{i}{2} \sin^2 \frac{i}{2} + \{1 + \cos 2(\theta + \theta')\} \sin^4 \frac{i}{2}\Big]$$
$$- n^2 a^4 \left[(2-2m)^2 A_1 \cos 2(\theta - \theta') + 4A_2 \cos 2\theta \right.$$
$$\left. + 4m^2 A_3 \cos 2\theta' + (2+2m)^2 A_4 \cos 2(\theta + \theta')\right].$$

Again, we have the equation

$$\frac{dH}{dt} = - rT,$$

which may be written

$$H \frac{dH}{dt} = -\frac{3}{2} n^3 m^2 a^4 \left[\sin 2 (\theta - \theta') \cos^4 \frac{i}{2} \right.$$

$$\left. + \sin 2\theta \cdot 2 \sin^2 \frac{i}{2} \cos^2 \frac{i}{2} + \sin 2 (\theta + \theta') \sin^4 \frac{i}{2} \right].$$

Substitute for μa its approximate value $n^2 a^4$ in the small terms; and we find from these two equations

$$- (1 - m) A_1 + 4 (1 - m)^3 A_1 + \frac{3}{2} m^2 (1 - m) \cos^4 \frac{i}{2} = -\frac{3}{2} m^2 \cos^4 \frac{i}{2},$$

$$- A_2 + 4 A_2 + \frac{3}{2} m^2 \cdot 2 \cos^2 \frac{i}{2} \sin^2 \frac{i}{2}$$

$$= -\frac{3}{2} m^2 \cdot 2 \cos^2 \frac{i}{2} \sin^2 \frac{i}{2},$$

$$- m A_3 + 4 m^3 A_3 + \frac{3}{2} m^3 \cdot 2 \cos^2 \frac{i}{2} \sin^2 \frac{i}{2} = 0,$$

$$- (1 + m) A_4 + 4 (1 + m)^3 A_4 + \frac{3}{2} m^2 (1 + m) \sin^4 \frac{i}{2} = -\frac{3}{2} m^2 \sin^4 \frac{i}{2}.$$

Therefore

$$A_1 = -\frac{3}{2} m^2 \cos^4 \frac{i}{2} \frac{2 - m}{(1 - m)(1 - 2m)(3 - 2m)},$$

$$A_2 = - 2 m^2 \cos^2 \frac{i}{2} \sin^2 \frac{i}{2},$$

$$A_3 = 3 m^2 \cos^2 \frac{i}{2} \sin^2 \frac{i}{2} \frac{1}{(1 - 2m)(1 + 2m)},$$

$$A_4 = -\frac{3}{2} m^3 \sin^4 \frac{i}{2} \frac{2 + m}{(1 + m)(1 + 2m)(3 + 2m)},$$

and

$$H^2 = n^2 a^4 \left[1 + m^2 - \frac{3}{2} m^2 \left(\cos^4 \frac{i}{2} + \sin^4 \frac{i}{2} \right) \right.$$

$$+ \frac{3}{2} \frac{m^2}{1 - m} \cos^4 \frac{i}{2} \cos 2 (\theta - \theta') + 3m \cos^2 \frac{i}{2} \sin^2 \frac{i}{2} \cos 2\theta$$

$$\left. + \frac{3}{2} \frac{m^2}{1 + m} \sin^4 \frac{i}{2} \cos 2 (\theta + \theta') \right];$$

where a is defined by
$$\mu = n^2 a^3.$$

If we preferred to define a so that the constant term in H^2 were equal to $n^2 a^4$, we should have

$$n^2 a^4 = \mu a + m^2 n^2 a^4 - \frac{3}{2} m^2 n^2 a^4 \left(\cos^4 \frac{i}{2} + \sin^4 \frac{i}{2} \right),$$

or $$\mu = n^2 a^3 \left[1 + \frac{1}{2} m^2 - 3m^2 \sin^2 \frac{i}{2} \cos^2 \frac{i}{2} \right].$$

Let us next find the latitude and the motion of the node.

Suppose that $i = i_0 + \Delta i,$
$$N = N_0 + \Delta N,$$

in which Δi, ΔN are small, i_0 is a constant, and N_0 varies slowly in proportion to the time, so that we may assume

$$\frac{dN_0}{dt} = -qn,$$

$$\Delta N = N_1 \sin 2 (\theta - \theta') + N_2 \sin 2\theta + N_3 \sin 2\theta' + N_4 \sin 2 (\theta + \theta'),$$
$$\Delta i = I_1 \cos 2 (\theta - \theta') + I_2 \cos 2\theta + I_3 \cos 2\theta' + I_4 \cos 2 (\theta + \theta').$$

Then remembering that

$$\frac{d\theta'}{dt} = mn - \frac{dN}{dt},$$

an expression that must be used in the terms of chief importance, we have

$$\frac{di}{dt} = \frac{d\Delta i}{dt} = -2 (1 - m) n I_1 \sin 2 (\theta - \theta') - 2n I_2 \sin 2\theta$$
$$- 2 \left(m - \frac{1}{n} \frac{dN}{dt} \right) n I_3 \sin 2\theta' - 2 (1 + m) n I_4 \sin 2 (\theta + \theta'),$$

$$\frac{dN}{dt} = \frac{dN_0}{dt} + \frac{d\Delta N}{dt} = -qn + 2(1-m)n N_1 \cos 2 (\theta - \theta') + 2n N_2 \cos 2\theta$$
$$+ 2 \left(m - \frac{1}{n} \frac{dN}{dt} \right) n N_3 \cos 2\theta' + 2 (1 + m) n N_4 \cos 2 (\theta + \theta')$$
$$+ \frac{dN_3}{di} \frac{di}{dt} \sin 2\theta',$$

in which the last term will be found to be required to get the constant q correctly to the order m^3.

These must be equated to

$$-\frac{Zr\cos\theta}{H}, \quad -\frac{Zr\sin\theta}{H\sin i}$$

respectively.

Hence

$$-q - 2mN_3{}^2 - mI_3\frac{dN_3}{di} = -\frac{3}{4}m^2\cos i;$$

therefore as a first approximation

$$q = \frac{3}{4}m^2\cos i;$$

hence

$$-2(1-m)I_1 = \frac{3}{4}m^2\sin i\cos^2\frac{i}{2}, \quad I_1 = -\frac{3}{8}\frac{m^2}{1-m}\sin i\cos^2\frac{i}{2},$$

$$-\quad 2I_2 = -\frac{3}{4}m^2\sin i\cos i, \quad I_2 = \frac{3}{8}m^2\sin i\cos i,$$

$$-2\left(m+\frac{3}{4}m^2\cos i\right)I_3 = -\frac{3}{4}m^2\sin i, \quad I_3 = \frac{3}{8}\frac{m\sin i}{1+\frac{3}{4}m\cos i},$$

$$-2(1+m)I_4 = -\frac{3}{4}m^2\sin i\sin^2\frac{i}{2}, \quad I_4 = \frac{3}{8}\frac{m^2}{1+m}\sin i\sin^2\frac{i}{2},$$

and

$$2(1-m)N_1 = -\frac{3}{4}m^2\cos^2\frac{i}{2}, \quad N_1 = -\frac{3}{8}\frac{m^2}{1-m}\cos^2\frac{i}{2},$$

$$2N_2 = \frac{3}{4}m^2\cos i, \quad N_2 = \frac{3}{8}m^2\cos i,$$

$$2\left(m+\frac{3}{4}m^2\cos i\right)N_3 = \frac{3}{4}m^2\cos i, \quad N_3 = \frac{3}{8}\frac{m\cos i}{1+\frac{3}{4}m\cos i},$$

$$2(1+m)N_4 = \frac{3}{4}m^2\sin^2\frac{i}{2}, \quad N_4 = \frac{3}{8}\frac{m^2}{1+m}\sin^2\frac{i}{2}.$$

Substitute above for the quantities I_3, N_3 and we get the second approximation to q,

$$q = \frac{3}{4}m^2\cos i - \frac{9}{32}m^2\cos^2 i + \frac{9}{64}m^3\sin^2 i.$$

It will be observed that I_3, N_3 are of lower order than the other coefficients, so that in order to obtain them correctly to the same order as the others we were obliged to retain small terms in $\dfrac{d\theta'}{dt}$ arising from the variability of N.

If we take the variable plane defined by the longitude of the node N_0 and the inclination i_0 as the plane to which the position of the Moon is referred, we have the latitude of the Moon above this plane

$$
= \quad \Delta i \sin\theta - \Delta N \sin i \cos\theta
$$

$$
= \quad \frac{3}{8} m \sin i \cos^2 \frac{i}{2} \left[\frac{1}{1 + \dfrac{3}{4} m \cos i} + \frac{m}{1 - m} \right] \sin(\theta - 2\theta')
$$

$$
- \frac{3}{8} m^2 \sin i \cos i \sin\theta
$$

$$
+ \frac{3}{8} m \sin i \sin^2 \frac{i}{2} \left[\frac{1}{1 + \dfrac{3}{4} m \cos i} - \frac{m}{1 + m} \right] \sin(\theta + 2\theta').
$$

LECTURE XVII.

ON HILL'S METHOD OF TREATING THE LUNAR THEORY.

LET us suppose the Moon to move in the plane of the ecliptic, and let us refer its motion to rectangular axes in rotation, the rotation being such that the axis of x passes always through the mean position of the Sun; that is, the axes rotate with angular velocity n', and if we suppose the Sun describes a circular orbit about the origin, its coordinates are

$$x' = a', \quad y' = 0.$$

Let x, y be the coordinates of the Moon.

Then the disturbing forces of the Sun upon the Moon relative to the Earth are

$$-\frac{m'}{\rho^2}\frac{x - a'}{\rho} - \frac{m'}{a'^2}, \quad -\frac{m'}{\rho^2}\frac{y}{\rho}$$

parallel to the axes of x and y respectively, where

$$\rho^2 = (x - a')^2 + y^2,$$

and the forces of the Earth on the Moon relative to the Earth are

$$-\frac{\mu}{r^2}\frac{x}{r}, \quad -\frac{\mu}{r^2}\frac{y}{r},$$

where $$r^2 = x^2 + y^2.$$

Now these forces may be written

$$\frac{d\Omega}{dx}, \quad \frac{d\Omega}{dy},$$

where $$\Omega = \frac{\mu}{r} + \frac{m'}{\rho} - \frac{m'x}{a'^2}.$$

But $\quad \dfrac{1}{\rho} = \dfrac{1}{a'} + \dfrac{x}{a'^2} + \dfrac{1}{a'^3}\left(x^2 - \dfrac{1}{2}y^2\right) + \dfrac{1}{a'^4}\left(x^3 - \dfrac{3}{2}xy^2\right) + \ldots$

Hence

$$\Omega = \dfrac{\mu}{r} + \dfrac{m'}{a'^3}\left(x^2 - \dfrac{1}{2}y^2\right) + \dfrac{m'}{a'^4}\left(x^3 - \dfrac{3}{2}xy^2\right) + \ldots$$

We have tacitly assumed the origin to be at the centre of the Earth; if we prefer to place it at the centre of gravity of the Earth and Moon, the necessary change is effected by multiplying the last terms, which correspond to the Parallactic Inequalities, by $(E - M)/(E + M)$.

Equating these forces to the accelerations of the Moon parallel to the coordinate axes, we have the equations of motion in the form

$$\dfrac{d^2x}{dt^2} - 2n'\dfrac{dy}{dt} - n'^2 x = \dfrac{d\Omega}{dx},$$

$$\dfrac{d^2y}{dt^2} + 2n'\dfrac{dx}{dt} - n'^2 y = \dfrac{d\Omega}{dy},$$

or, as they may be written,

$$\dfrac{d^2x}{dt^2} - 2n'\dfrac{dy}{dt} = \dfrac{dR}{dx},$$

$$\dfrac{d^2y}{dt^2} + 2n'\dfrac{dx}{dt} = \dfrac{dR}{dy},$$

where $\quad R = \Omega + \dfrac{1}{2}n'^2(x^2 + y^2)$

$$= \dfrac{\mu}{r} + \dfrac{3}{2}n'^2 x^2 + \dfrac{n'^2}{a'}\left(x^3 - \dfrac{3}{2}xy^2\right) + \ldots$$

Now suppose we have found values of x and y which satisfy this pair of equations and which involve two arbitrary constants. This may be accomplished by taking assumed developments

$$x = \Sigma a_i \cos i\,(t + \gamma),$$

$$y = \Sigma b_i \sin i\,(t + \gamma),$$

substituting in the equations, and equating coefficients of the various terms. The solution found will include the Variation and the Parallactic Inequalities. Let it be required to amend this solution by the introduction of the remaining two arbitrary constants that are required for a complete solution.

Let the additional terms that we seek be δx, δy, which we shall suppose so small that their squares and products may be neglected, let us consider first the terms which are multiplied by the first power of one of the new arbitraries, the original particular solution corresponding to the case in which this arbitrary is zero.

Then δx, δy are determined by the equations

$$\frac{d^2 \delta x}{dt^2} - 2n' \frac{d\delta y}{dt} = \frac{d^2 R}{dx^2} \delta x + \frac{d^2 R}{dx\,dy} \delta y + X,$$

$$\frac{d^2 \delta y}{dt^2} + 2n' \frac{d\delta x}{dt} = \frac{d^2 R}{dx\,dy} \delta x + \frac{d^2 R}{dy^2} \delta y + Y,$$

where X, Y are supposed known functions of x, y or of t, and have been added here to include disturbing causes not allowed for in the above form of R.

Multiply the original equations by $\frac{dx}{dt}$, $\frac{dy}{dt}$ and add:

$$\frac{d^2 x}{dt^2} \frac{dx}{dt} + \frac{d^2 y}{dt^2} \frac{dy}{dt} = \frac{dR}{dx} \frac{dx}{dt} + \frac{dR}{dy} \frac{dy}{dt}$$

$$= \frac{dR}{dt},$$

since x, y are the only functions of t that R involves; whence

$$\left(\frac{dx}{dt}\right)^2 + \left(\frac{dy}{dt}\right)^2 = 2R + C,$$

where C is an arbitrary constant; this is the integral known as Jacobi's Integral.

Let us write

$$\frac{dx}{dt} = V \cos \phi, \qquad \frac{dy}{dt} = V \sin \phi;$$

then we have $\qquad\qquad V^2 = 2R + C;$

and from the original equations themselves

$$\frac{dV}{dt} = \frac{d^2 x}{dt^2} \cos \phi + \frac{d^2 y}{dt^2} \sin \phi$$

$$= \left(\frac{d^2 x}{dt^2} - 2n' \frac{dy}{dt}\right) \cos \phi + \left(\frac{d^2 y}{dt^2} + 2n' \frac{dx}{dt}\right) \sin \phi$$

$$= \frac{dR}{dx} \cos \phi + \frac{dR}{dy} \sin \phi;$$

and

$$V \frac{d\phi}{dt} = -\frac{d^2x}{dt^2} \sin\phi + \frac{d^2y}{dt^2} \cos\phi$$

$$= -\left(\frac{d^2x}{dt^2} - 2n'\frac{dy}{dt}\right) \sin\phi + \left(\frac{d^2y}{dt^2} + 2n'\frac{dx}{dt}\right) \cos\phi$$

$$+ 2n'\left(\frac{dx}{dt}\cos\phi + \frac{dy}{dt}\sin\phi\right),$$

or

$$V\left(\frac{d\phi}{dt} + 2n'\right) = -\frac{dR}{dx}\sin\phi + \frac{dR}{dy}\cos\phi.$$

And from these, differentiating and substituting for $\frac{dx}{dt}$, $\frac{dy}{dt}$, we get

$$\frac{d^2V}{dt^2} - V\frac{d\phi}{dt}\left(\frac{d\phi}{dt} + 2n'\right) = V\left[\begin{array}{l} \frac{d^2R}{dx^2}\cos^2\phi \\ \\ + 2\frac{d^2R}{dxdy}\cos\phi\sin\phi + \frac{d^2R}{dy^2}\sin^2\phi \end{array}\right],$$

$$V\frac{d^2\phi}{dt^2} + 2\frac{dV}{dt}\left(\frac{d\phi}{dt} + n'\right) = V\left[\begin{array}{l} -\frac{d^2R}{dx^2}\sin\phi\cos\phi \\ \\ + \frac{d^2R}{dxdy}(\cos^2\phi - \sin^2\phi) + \frac{d^2R}{dy^2}\sin\phi\cos\phi \end{array}\right].$$

LECTURE XVIII.

ON HILL'S METHOD OF TREATING THE LUNAR THEORY (*continued*).

THE equations for δx, δy are

$$\frac{d^2 \delta x}{dt^2} - 2n' \frac{d\delta y}{dt} = \frac{d^2 R}{dx^2} \delta x \; + \frac{d^2 R}{dx\,dy} \delta y + X,$$

$$\frac{d^2 \delta y}{dt^2} + 2n' \frac{d\delta x}{dt} = \frac{d^2 R}{dx\,dy} \delta x + \frac{d^2 R}{dy^2} \; \delta y + Y;$$

the equations for x, y are

$$\frac{d^2 x}{dt^2} - 2n' \frac{dy}{dt} = \frac{dR}{dx},$$

$$\frac{d^2 y}{dt^2} + 2n' \frac{dx}{dt} = \frac{dR}{dy}.$$

Multiply the former pair by $\dfrac{dx}{dt}$, $\dfrac{dy}{dt}$ respectively, and the latter pair by $\dfrac{d\delta x}{dt}$, $\dfrac{d\delta y}{dt}$, and add all together; we get

$$\frac{dx}{dt} \frac{d^2 \delta x}{dt^2} + \frac{d^2 x}{dt^2} \frac{d\delta x}{dt} + \frac{dy}{dt} \frac{d^2 \delta y}{dt^2} + \frac{d^2 y}{dt^2} \frac{d\delta y}{dt}$$

$$= \left(\frac{d^2 R}{dx^2} \frac{dx}{dt} + \frac{d^2 R}{dx\,dy} \frac{dy}{dt} \right) \delta x + \left(\frac{d^2 R}{dx\,dy} \frac{dx}{dt} + \frac{d^2 R}{dy^2} \frac{dy}{dt} \right) \delta y$$

$$+ \frac{dR}{dx} \frac{d\delta x}{dt} + \frac{dR}{dy} \frac{d\delta y}{dt} + X \frac{dx}{dt} + Y \frac{dy}{dt}.$$

Now
$$\frac{d^2 R}{dx^2} \frac{dx}{dt} + \frac{d^2 R}{dx\,dy} \frac{dy}{dt} = \frac{d}{dt} \frac{dR}{dx},$$

$$\frac{d^2 R}{dx\,dy} \frac{dx}{dt} + \frac{d^2 R}{dy^2} \frac{dy}{dt} = \frac{d}{dt} \frac{dR}{dy}.$$

Then our equation may be integrated

$$\frac{dx}{dt}\frac{d\,\delta x}{dt} + \frac{dy}{dt}\frac{d\,\delta y}{dt} = \frac{dR}{dx}\,\delta x + \frac{dR}{dy}\,\delta y + T,$$

where

$$T = \int \left(X\,\frac{dx}{dt} + Y\,\frac{dy}{dt} \right) dt,$$

so that T is a known function of t, which involves an arbitrary constant.

Now let us assume

$$\delta x = v\cos\phi - w\sin\phi,$$
$$\delta y = v\sin\phi + w\cos\phi.$$

Substitute above for $\dfrac{dx}{dt},\ \dfrac{dy}{dt},\ \dfrac{d\,\delta x}{dt},\ \dfrac{d\,\delta y}{dt}$; we find

$$V\left(\frac{dv}{dt} - w\frac{d\phi}{dt}\right) = \left(\frac{dR}{dx}\cos\phi + \frac{dR}{dy}\sin\phi\right) v$$
$$+ \left(-\frac{dR}{dx}\sin\phi + \frac{dR}{dy}\cos\phi\right) w + T.$$

But

$$\frac{dR}{dx}\cos\phi + \frac{dR}{dy}\sin\phi = \frac{dV}{dt},$$

$$-\frac{dR}{dx}\sin\phi + \frac{dR}{dy}\cos\phi = V\left(\frac{d\phi}{dt} + 2n'\right).$$

Therefore $\quad V\left(\dfrac{dv}{dt} - w\dfrac{d\phi}{dt}\right) = \dfrac{dV}{dt}\,v + V\left(\dfrac{d\phi}{dt} + 2n'\right) w + T,$

or $\qquad V\,\dfrac{dv}{dt} - \dfrac{dV}{dt}\,v = \qquad 2wV\left(\dfrac{d\phi}{dt} + n'\right) \qquad + T,$

whence $\qquad \dfrac{v}{V} = \int \dfrac{2}{V}\left(\dfrac{d\phi}{dt} + n'\right) w\,dt + \int \dfrac{T}{V^2}\,dt.$

An arbitrary constant is included on the right. This equation shews that when w is known, v can be found; it remains to determine w.

Now by actual differentiation

$$\cos\phi\,\frac{d\,\delta x}{dt} + \sin\phi\,\frac{d\,\delta y}{dt} = \frac{dv}{dt} - w\frac{d\phi}{dt}$$

$$-\sin\phi\,\frac{d^2\delta x}{dt^2} + \cos\phi\,\frac{d^2\delta y}{dt^2} = \frac{d^2w}{dt^2} + 2\frac{dv}{dt}\frac{d\phi}{dt} - w\left(\frac{d\phi}{dt}\right)^2 + v\frac{d^2\phi}{dt^2}.$$

Also multiplying the differential equations for δx, δy by $-\sin\phi$, $\cos\phi$, respectively and adding

$$-\sin\phi\,\frac{d^2\delta x}{dt^2}+\cos\phi\,\frac{d^2\delta y}{dt^2}+2n'\left(\cos\phi\,\frac{d\delta x}{dt}+\sin\phi\,\frac{d\delta y}{dt}\right)$$

$$=\left(-\frac{d^2R}{dx^2}\sin\phi+\frac{d^2R}{dxdy}\cos\phi\right)\delta x+\left(-\frac{d^2R}{dxdy}\sin\phi+\frac{d^2R}{dy^2}\cos\phi\right)\delta y$$

$$-X\sin\phi+Y\cos\phi.$$

Substitute and we find

$$\frac{d^2w}{dt^2}+2\frac{dv}{dt}\frac{d\phi}{dt}-w\left(\frac{d\phi}{dt}\right)^2+v\frac{d^2\phi}{dt^2}+2n'\left(\frac{dv}{dt}-w\frac{d\phi}{dt}\right)$$

$$=\quad v\left[-\frac{d^2R}{dx^2}\sin\phi\cos\phi+\frac{d^2R}{dxdy}(\cos^2\phi-\sin^2\phi)+\frac{d^2R}{dy^2}\sin\phi\cos\phi\right]$$

$$+w\left[\frac{d^2R}{dx^2}\sin^2\phi\quad-2\frac{d^2R}{dxdy}\sin\phi\cos\phi\quad+\frac{d^2R}{dy^2}\cos^2\phi\right]$$

$$-X\sin\phi+Y\cos\phi.$$

Now we have seen

$$\frac{dv}{dt}=\frac{v}{V}\frac{dV}{dt}+2\left(\frac{d\phi}{dt}+n'\right)w+\frac{T}{V}.$$

Substitute for $2\left(\dfrac{d\phi}{dt}+n'\right)\dfrac{dv}{dt}$ on the left.

We get

$$\frac{d^2w}{dt^2}+v\left[\frac{d^2\phi}{dt^2}+\frac{2}{V}\frac{dV}{dt}\left(\frac{d\phi}{dt}+n'\right)\right]+w\left[4\left(\frac{d\phi}{dt}+n'\right)^2-\left(\frac{d\phi}{dt}\right)^2-2n'\frac{d\phi}{dt}\right]$$

$$+2\left(\frac{d\phi}{dt}+n'\right)\frac{T}{V}$$

$$=\quad v\left[-\frac{d^2R}{dx^2}\sin\phi\cos\phi+\frac{d^2R}{dxdy}(\cos^2\phi-\sin^2\phi)+\frac{d^2R}{dy^2}\sin\phi\cos\phi\right]$$

$$+w\left[\frac{d^2R}{dx^2}\sin^2\phi\quad-2\frac{d^2R}{dxdy}\sin\phi\cos\phi\quad+\frac{d^2R}{dy^2}\cos^2\phi\right]$$

$$-X\sin\phi+Y\cos\phi.$$

But by the equations proved at the end of Lecture XVII, the terms in v cancel one another, and we are left with the equation for w:

$$\frac{d^2w}{dt^2} + w \left[3 \left(\frac{d\phi}{dt} \right)^2 + 6n' \frac{d\phi}{dt} + 4n'^2 - \frac{d^2R}{dx^2} \sin^2 \phi \right.$$

$$\left. + 2 \frac{d^2R}{dx\,dy} \sin \phi \cos \phi - \frac{d^2R}{dy^2} \cos^2 \phi \right]$$

$$= - 2 \left(\frac{d\phi}{dt} + n' \right) \frac{T}{V} - X \sin \phi + Y \cos \phi.$$

Or since

$$\cos \phi = \frac{1}{V} \frac{dx}{dt}, \qquad \sin \phi = \frac{1}{V} \frac{dy}{dt}$$

$$\frac{d\phi}{dt} + 2n' = \frac{1}{V} \left(-\frac{dR}{dx} \sin \phi + \frac{dR}{dy} \cos \phi \right)$$

the coefficient of w is

$$\frac{3}{V^4} \left(-\frac{dR}{dx} \frac{dy}{dt} + \frac{dR}{dy} \frac{dx}{dt} \right)^2 - 6 \frac{n'}{V^2} \left(-\frac{dR}{dx} \frac{dy}{dt} + \frac{dR}{dy} \frac{dx}{dt} \right) + 4n'^2$$

$$- \frac{1}{V^2} \left\{ \frac{d^2R}{dx^2} \left(\frac{dy}{dt} \right)^2 - 2 \frac{d^2R}{dx\,dy} \frac{dx}{dt} \frac{dy}{dt} + \frac{d^2R}{dy^2} \left(\frac{dx}{dt} \right)^2 \right\}$$

$$= P, \text{ say.}$$

This function P is a known function of t; it may be seen that if

$$x = \Sigma a_i \cos i\,(t + \gamma),$$

$$y = \Sigma b_i \sin i\,(t + \gamma),$$

then P may be developed in the form

$$P = \Sigma A_i \cos i\,(t + \gamma).$$

Hence if we omit the terms X, Y, due to other disturbances not yet allowed for, the equation for w assumes the form

$$\frac{d^2w}{dt^2} + w \left[A_0 + A_1 \cos (t + \gamma) + A_2 \cos 2\,(t + \gamma) + \dots \right] = 0.$$

This is identical in form with the equation treated in Lecture XIV, to find the motion of the node. The value of w may be found by the method there employed, and the value of v deduced from it.

Printed in the United States
By Bookmasters